MW00607797

Focus On

Grades 5-8

MIDDLE SCHOOL

3rd Edition

Rebecca W. Keller, PhD

Real Science-4-Kids

Illustrations: Janet Moneymaker

Focus On Middle School Geology Student Textbook—3rd Edition (hardcover)
ISBN 978-1-941181-71-3

Published by Gravitas Publications Inc.
www.gravitaspublications.com
www.realscience4kids.com

Contents

Chapter 1 What Is Geology?

1.1 Introduction

Most people probably don't often think about Earth being the place where they live. When you ask someone where they live, they might reply "on 4th Street" or "in Minneapolis," but rarely do you hear "on Earth." In fact, everyone lives on Earth, and as far as we know, there is no one living on any other planet. Most people don't often wonder about what the Earth is made of or think about Earth being only one of many planets in the universe.

So what is the Earth? What is it made of? Has it always been this way, or has it changed? Why can Earth support life and the Moon can't? What makes Earth special? Finding out about the Earth, what it is made of, and how it changes are inquiries into the scientific field of geology.

1.2 What Is Geology?

The word geology comes from the Greek root words *geo* which means "earth" and *logy* which means "the study of." So geology is "the study of Earth." Geology is a science that focuses on bringing about a better understanding of the structure and history of Earth, the planet we live on.

The field of geology is divided into two broad categories — physical geology and historical geology.

Physical geology examines the chemical and physical nature of Earth and also the processes that operate on and inside Earth. Physical geology explores the Earth's surface, the processes that form the Earth, and the heat energy that drives these processes.

Historical geology examines the origin of Earth and incorporates biology, chemistry, and physics in an attempt to create a chronological narrative, or story, about how the Earth came into being and how it has changed over time.

1.3 Interpreting Geological Data

Science has two parts. One part is collecting scientific data through observation and experimentation. The second part is to find out what the data mean, and this is called interpretation. Interpreting scientific data is the process that scientists use to draw conclusions, formulate theories, and develop scientific laws and principles.

Because science is a human endeavor, the interpretations of scientific data are subject to human bias and presupposition. In science a presupposition is an assumption about how something works and is usually based on preexisting beliefs and sometimes on previous experience. For example, because it is known that planes can fly, if a plane is seen in a hangar, the assumption may be that the plane will fly even if this particular plane has not been seen in the air.

It is not incorrect for presuppositions to be used in science, and scientists do use them all the time. However, scientists may begin research with differing presuppositions, and even though scientists use logic and strive to be objective, there is often disagreement about how scientific data should be interpreted.

Disagreements in science are a vital part of scientific investigation and should be encouraged because they can lead to new ideas and new ways of thinking about observations. However, many people, including scientists, are uncomfortable with arguing.

Many scientists see the world in a certain way, and since they are not open to other points of view, they insist that every other scientist see the world in the same way. The way someone "sees the world" is called their worldview. Someone's worldview is made up of the philosophies and beliefs that they use to understand the world around them. There are as many different worldviews as there are people because no two people see the world in exactly the same way. This difference between worldviews causes many arguments in the scientific community.

Geologists with different worldviews disagree about how the Earth came into being, how old it is, and how it has changed over time. Most of these disagreements occur in the area of historical geology and the historical narrative for Earth. But sometimes they also occur in the area of physical geology, especially if historical presuppositions are used to develop physical theories.

This text will focus on physical geology and will not discuss different historical narratives for Earth. However, it is important to keep in mind that interpreting geological data is an exciting and dynamic part of studying geology and that disagreements help advance our understanding of Earth.

1.4 Why Study Earth?

Have you ever wondered why mountains are very tall and oceans are deep? Have you ever wondered why a desert has very little rain but it rains all the time in a tropical forest? Have you ever wondered where earthquakes come from or why some mountains erupt as volcanoes and others don't? Have you ever thought about where we get iron, copper, and oil?

Earth is a unique planet in our solar system and provides the habitat for all living things, including human beings! By studying Earth we begin to understand what makes Earth so special.

We can learn where to find natural resources that improve the quality of our lives. We can also learn what causes certain geological catastrophes, such as earthquakes and hurricanes so we can help people prepare for and protect against devastating losses. We can learn how beautiful landscapes or vast forests have developed and work to preserve Earth's geological features. By studying Earth we can both protect Earth's native beauty and resources and use them to provide a future for the next generation.

1.5 What Do Geologists Study?

If you've ever met a geologist, you might have noticed that they can spend a lot of time outdoors — hiking up mountains and walking through fields. They also tend to collect a lot of rocks!

Geologists study the Earth, and in order to do this they go outside to explore and observe what is on and in the Earth.

A geologist studies snow

Photo Credit: US Geological Survey/by Lyn Topinka

There is much to investigate since Earth is a complex planet that is changing every moment of every day — whether it's rocks falling in a landslide, the top of a mountain breaking apart as a volcano erupts, or the ground moving in an earthquake.

A geologist studies erosion near the Dead Sea, Israel

Photo Credit: Mark A. Wilson, The College of Wooster

There are many different branches of geology, and each focuses on different aspects of the Earth. One branch of geology is called geochemistry. Geochemistry is the study of the chemistry of Earth. The Earth is made of atoms and molecules, just like all matter, and geochemists study the specific types of atoms and molecules that form Earth. In order to study the chemistry of Earth, geochemists take samples of the rocks, minerals, soils, and other matter that Earth is made from and analyze the samples. In this book we will take a close look at the matter that makes up Earth.

Another branch of geology is called structural geology, which deals with the internal structure, form, and arrangement of rocks. Structural geologists study how rocks deform to make mountains and valleys. Deformation occurs when the shape and size of rocks change due to bending, twisting, or fracturing.

There are also geologists who look for energy resources like gas, oil, and coal. This branch of geology is called resource geology. Resource geologists look for the natural resources that humans need for living on Earth.

Deformed rock near Barstow, California

Photo Credit: Mark A. Wilson, The College of Wooster

When humans interact with Earth, sometimes they modify or change the environment, which includes those factors that affect living organisms, such as landscape, water, and air quality. Geologists need to monitor changes to the environment to help protect and clean it up. Geologists who study environmental changes caused by human activities are called environmental geologists.

1.6 Geology and the Scientific Method

In the science of geology, Earth itself is often the laboratory. Because things can happen very slowly on Earth, designing experiments and collecting results can be difficult. For this reason, geologists address many questions about Earth by observing features in rocks or landscapes, collecting rock samples, and using electronic equipment, among other things. Geologists formulate hypotheses based on what they've observed and then test those hypotheses. The conclusions they draw may be modified if the hypotheses are found to be incorrect.

Because many of Earth's features are not as testable or as provable as experiments in chemistry or physics, there is often disagreement about what the data mean. Even so, general principles about Earth's features can be proposed. These principles can be thought of as "scientific maps," and it is useful to keep in mind that scientific maps can change with new data or revised hypotheses.

1.7 Summary

- Geology is the study of Earth.
- Geology is divided into two broad categories — physical geology and historical geology.

- Physical geology examines the chemistry and physics of Earth. Historical geology attempts to create a chronological and historical narrative about Earth's origins.
- A person's worldview is made up of the philosophies and beliefs they use to understand the world around them.
- Geologists study Earth to learn more about Earth's features, to protect Earth's environment, and to find Earth's resources.

1.8 Some Things to Think About

- Go outside and look around. Based on what you see, describe what you think the Earth is made of.
- Which category of geology do you think would be the most interesting to study: physical geology or historical geology? Why?
- Do you think it can be difficult for people to listen to others who are expressing a very different worldview? Why or why not?
- Do you think it can be difficult for people to change their worldview? Why or why not?
- What would you most like to learn about Earth? Why would you find this interesting?
- Which of the following branches of geology do you think would be the most interesting to study? Why?

  Geochemistry
  Structural geology
  Resource geology
  Environmental geology
- How would you define a "scientific map"?

Chapter 2 Technology in Geology

2.1 Introduction

Geologists use a variety of tools to study Earth. A hundred years ago geologists had only a few tools available. This limited what early geologists were able to study. But today, modern geologists use many different tools to measure, dig, monitor, and explore the Earth. Today's modern geologists use hand tools, electronic tools, and a variety of other tools.

2.2 Hand Tools

Rock Hammer

Courtesy of Mark A. Wilson, The College of Wooster

Probably the most common hand tool used by geologists is the rock hammer or handpick. A rock hammer is a lightweight hammer used to break apart rocks or pry loose rock and soil away from other rocks. A rock hammer usually has a blunt end for hammering and a chiseled end for prying. Rock hammers vary in size and weight but usually are small enough and light enough to carry in a backpack. Sometimes when geologists photograph a rock sample, they will indicate the size of the sample by putting their hammer in the photo.

Another type of hand tool used by geologists is the crack hammer. A crack hammer is a small sledgehammer with two blunt ends, and it is heavier than a rock hammer. Crack hammers are used to break apart larger rocks.

The compass is another hand tool often used by geologists. In order to observe Earth's features, geologists must often travel outdoors in undeveloped wilderness areas. A compass can help a geologist navigate the terrain and find the road back to civilization after a long hike!

Along with a compass, geologists commonly use several different types of maps. Using different types of maps is a good way to understand a region.

City and road maps show cities, city landmarks, and roads and can be used to study geological features within and between populated areas.

Topographic maps give geologists an idea of how high or low an area might be. A topographic map has contour lines that show the shape and elevation of an area. When the lines are close together, it means the area is steep. When the lines are far apart, it means the area is flat or less steep.

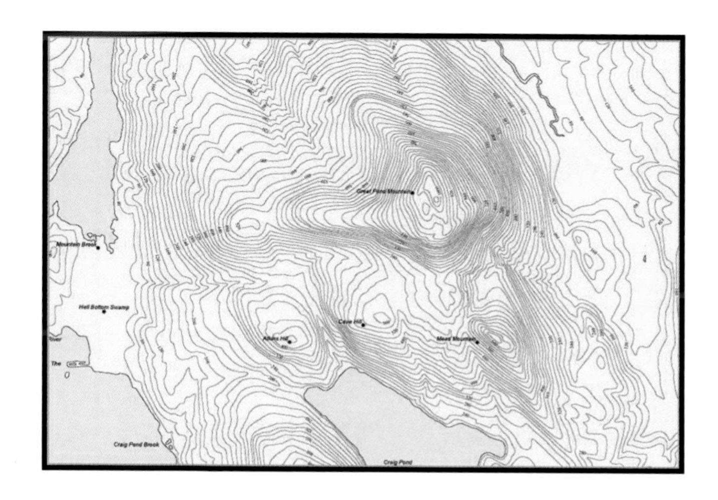

Climate maps give information about the climate of a particular region. A climate map shows the amount of precipitation, or how much rain or snow an area receives.

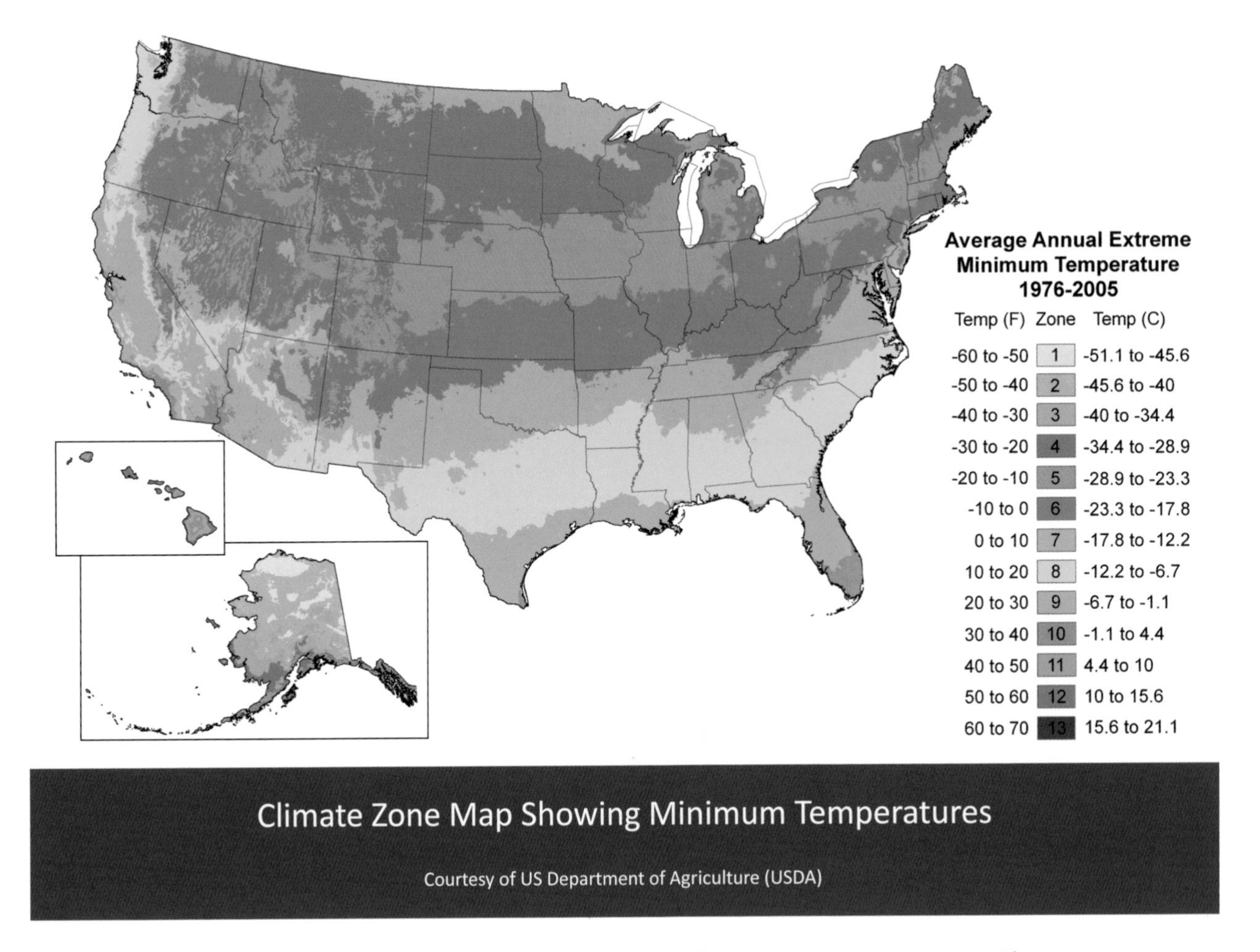

Climate Zone Map Showing Minimum Temperatures

Courtesy of US Department of Agriculture (USDA)

Climate maps also show differences in temperature from region to region. Climate maps are divided into climate zones accompanied by a key that shows which areas are dry or moist, hot or cold.

2.3 Electronic Tools

There are many modern electronic tools geologists can use to study the Earth. For example, imagine going into a dense forest, far from cities and roads, to study a particular type of rock or geological formation. If you are unfamiliar with the area, and if the maps are not very reliable, it might be easy to get lost. A modern tool you can use to find your way is a GPS or global positioning system. A GPS is a device that uses satellite information to provide the location and time on or near the Earth. A GPS calculates position by precisely timing signals sent by satellites high above the Earth. The satellite sends messages to the GPS, and based on these signals, the GPS can calculate an exact location.

A GPS is handy for getting to places above ground, but what if you wanted to know what is below ground? One way to explore features below ground is to dig a hole, but holes can be dug only so far and digging can disrupt critical features a geologist might want to study. To "see" below the surface of the Earth, geologists can now use GPR, or ground penetrating radar. GPR is a method that uses high frequency radio waves to image below the surface. High frequency radio waves can be transmitted into the ground, and when the radio waves encounter an object, the waves will be reflected. A computer collects the reflected waves and creates an image of the objects below. GPR can be used to study ice formations, soils, groundwater, and even dinosaur bones!

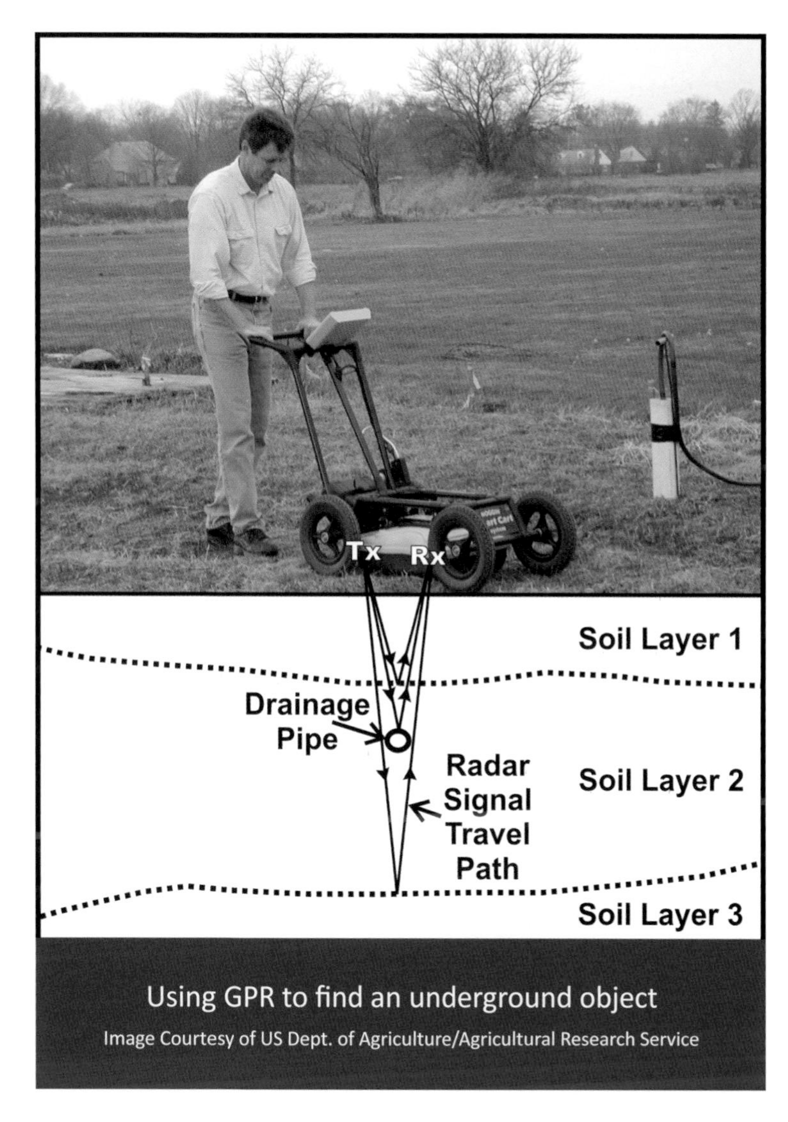

Using GPR to find an underground object
Image Courtesy of US Dept. of Agriculture/Agricultural Research Service

A GPR is useful for observing static objects — those that stay the same. But what about observing dynamic processes — those that change over time, like earthquakes and volcanic eruptions? To observe dynamic processes, geologists can use a seismometer or a seismograph. Both of these instruments measure motions in the ground. The words seismometer and seismograph come from the Greek word *seismos* which means "to shake or quake."

Geologist with seismographs
Courtesy of US Geological Survey (USGS)/by J. K. Nakata

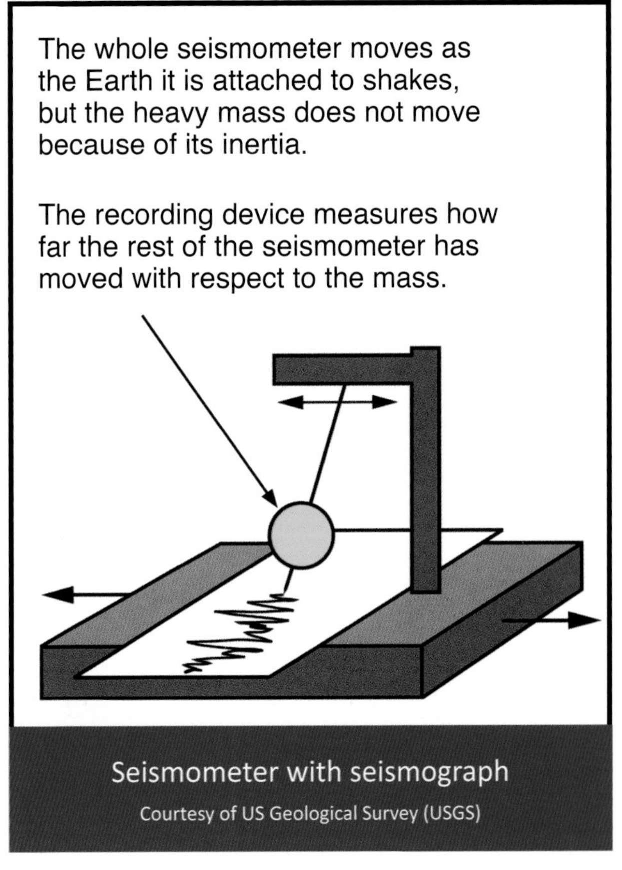

Seismometer with seismograph
Courtesy of US Geological Survey (USGS)

A seismometer is an instrument used "to measure shakes and quakes" and seismograph means "to draw shakes and quakes." Both instruments detect and measure the motion of the Earth's surface.

2.4 Other Tools

Drill bit used to take core samples of rock
Courtesy of US Geological Survey (USGS)

Sometimes a geologist might want to take a sample of rock from deep inside the Earth's surface. To do this a core drill can be used. A core drill is a drill specifically designed to go several hundreds to several thousands of feet into the Earth's surface. The core drill cuts a circular hole as it moves downward, creating a cylinder of rock, or core sample, that can be pulled out of the Earth. Core drills can be used for mineral exploration and to study the layers of rock deep in the Earth.

A rock and mineral test kit is another useful tool for geologists. Once a rock sample is collected, the rock and mineral test kit can be used to test for color, hardness, and how a sample reacts with certain acids. All of these features are helpful in identifying the types of minerals in the sample. A rock and mineral test kit is very useful for field geologists because it is small enough to be carried in a backpack.

2.5 Satellites

Satellites have become essential data gathering tools for geologists. In space, a satellite is an object that orbits another object. For example, the Moon is a satellite of Earth because it orbits Earth. An artificial satellite is a machine that has been launched into space and put into orbit around Earth. To gather information about Earth, the United States and many other countries have satellites in orbit that collect data about such features as the atmosphere, climate, geodynamics, gravity, weather, the oceans, ice, groundwater, the Sun and its influence on Earth, and the magnetosphere.

In 1972 NASA launched the first of what are now called the Landsat satellites. Using over 40 years of this satellite imagery, scientists are able to track how Earth's surface changes both slowly and quickly over time. Natural events such as hurricanes, blizzards, volcanic eruptions, earthquakes, and forest fires can be studied. The information collected can be used to identify the effects of natural events on Earth's surface and how these events affect people and the environment. The impact of human caused changes to the Earth, such as pollution, population growth, and deforestation, can also be studied, leading to new ideas for ways to protect the environment.

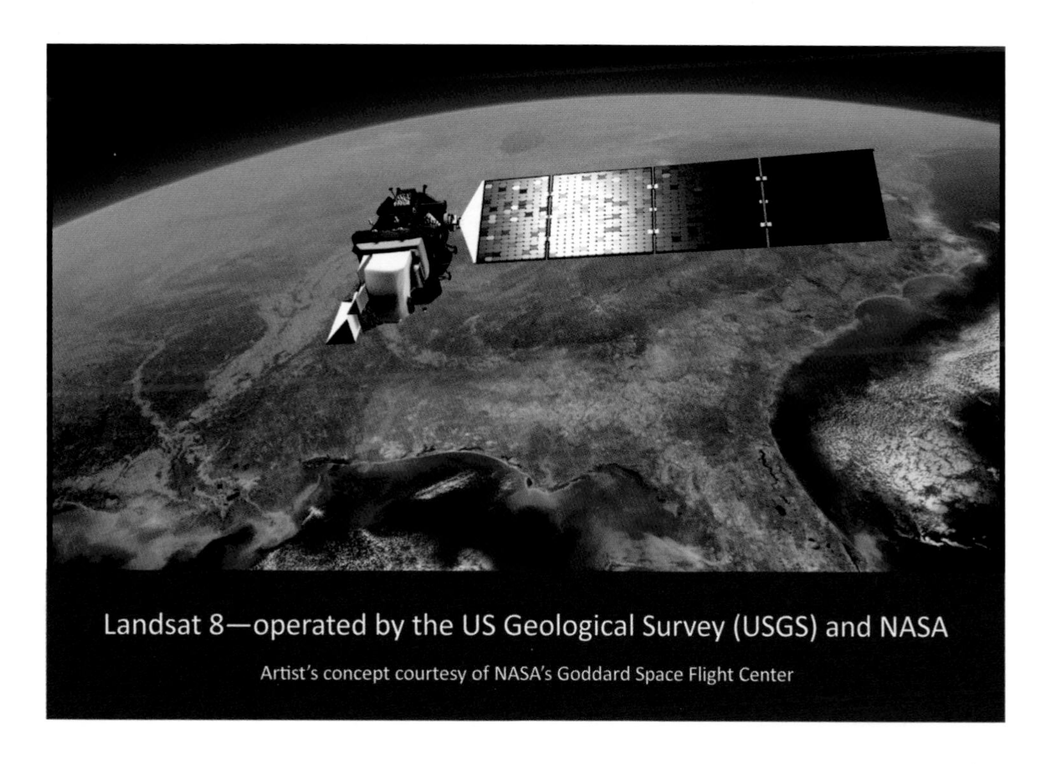

Landsat 8—operated by the US Geological Survey (USGS) and NASA

Artist's concept courtesy of NASA's Goddard Space Flight Center

The SMAP (Soil Moisture Active Passive) satellite launched by NASA in 2015 measures and maps the amount of water in the top 5 cm (2 inches) of soil around the world. This will help scientists better understand the water cycle and how storms, drought, and the changes of season affect moisture in the soil, flooding, and the growing of crops. It will also provide data for studies of weather, climate, and the carbon cycle.

The SMAP satellite antenna scans a swath 1000 km (620 mi.) wide as it rotates, collecting data and mapping the entire surface of Earth in 2-3 days.

Artists concept courtesy of NASA/JPL-Caltech

2.6 Summary

- Geologists use many different tools to measure, dig, monitor, and explore the Earth.
- Rock hammers and crack hammers are hand tools geologists use to collect rock samples.
- Geologists use various types of maps, such as topographic maps and climate maps, to help them understand the region of the Earth being studied.
- Electronic tools are helpful in finding locations, measuring vibrations in the ground, and visualizing objects below the surface of the Earth.
- Satellites are important tools for gathering data about Earth.

2.7 Some Things to Think About

- Do you think today's geologists ever use horses and mules when they are out in the field? Why or why not?
- If you were a geologist out hiking and looking at rock samples, what tools would you want to have with you? Why would you want each of these?

- If you were a farmer trying to decide which crops would grow well in your area, what kind of map would you look at? What would it tell you?
- When would you use a GPS device?
- What do you think you might discover by using a GPR device?
- What do you think you might discover by using a seismometer and a seismograph?
- If you could take a core sample from deep in the Earth, what do you think you would find?
- If you looked at Landsat images of your area, what do you think you could discover?
- What do you think you could find out about your area from SMAP data?

Chapter 3 Rocks, Minerals, and Soils

3.1 Introduction

When you walk outside or sit on the ground, you might wonder: What exactly is the Earth made of? If you dig your fingers into the ground you will discover that the ground is full of dirt (soil) and rocks. If you look under your house or under the library, you will find that rocks and dirt are also underneath the buildings. If you go to the tallest mountain or dive to the deepest part of the ocean, you will still find soil and rocks. In some places the soil is soft or sandy, and you can grab a handful of it and roll it around in your hand, letting it flow through your fingers. In other places the soil may be so full of big rocks that you cannot grab it with your fingers but instead need a jackhammer or drill to dig deeper.

The ground we walk on and build our houses and roads on is the outermost layer of the Earth and is called the crust. (We will learn more about the different layers of the Earth in the next chapter.) The crust is composed of rocks and soils that are made of different chemical elements.

Different rocks are made from different minerals which are made from different chemical elements combined in a variety of ways. Different soils are made from different organic and inorganic materials. Both rocks and soils form the Earth.

3.2 Minerals

Minerals are the building blocks of rocks and soils. There are almost 4,000 different types of minerals. However, only a few dozen of those minerals are rock-forming minerals. There are eight chemical elements that make up the majority of rock-forming minerals: oxygen, silicon, aluminum, iron, calcium, sodium, potassium, and magnesium.

What is a mineral? If you look in your backyard and see plastic toys, trees, water, ice, and rocks, how can you tell which of these are minerals and which are not minerals? In geology, in order for a material to be considered a mineral it must meet the following criteria:

1) It must occur naturally.
2) It must be a solid.
3) It must be inorganic, meaning it does not contain the element carbon.
4) It must have an internal structure, or organization, of the atoms.

Plastic toys are not minerals because they are not naturally formed. Trees are not minerals because even though they are naturally formed, they contain carbon (are organic). Water is not a mineral because even though it is naturally formed and does not contain carbon, it is not a solid. Ice is not a mineral because even though it is naturally formed, a solid, and inorganic, it does not have an internal structure or organization of the atoms.

There are many different kinds of minerals found in rocks and soils; however, the majority of rock-forming minerals are silicates. Silicates contain the chemical compound silicate, which is an oxygen-silicon compound that has four oxygens bonded to one silicon in a tetrahedral shape. (A tetrahedron is a solid geometric shape that has four sides.)

In a silicate molecule, the larger oxygen atoms surround the smaller central silicon atom to form a tetrahedral unit.

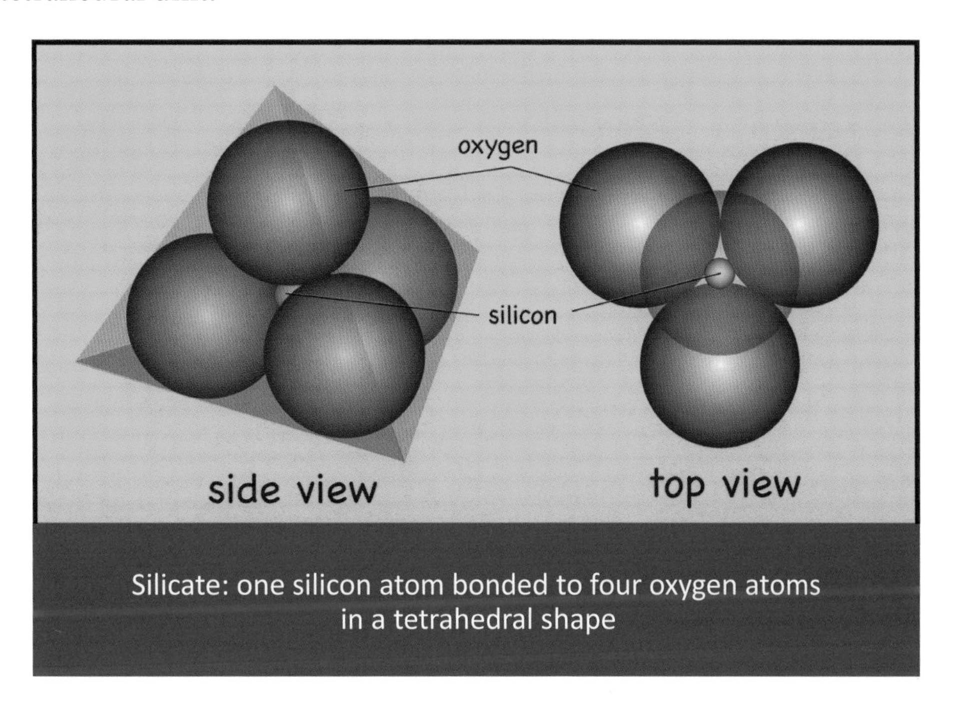

Silicate: one silicon atom bonded to four oxygen atoms in a tetrahedral shape

These units join together to make single and double chains of molecules and also sheet structures.

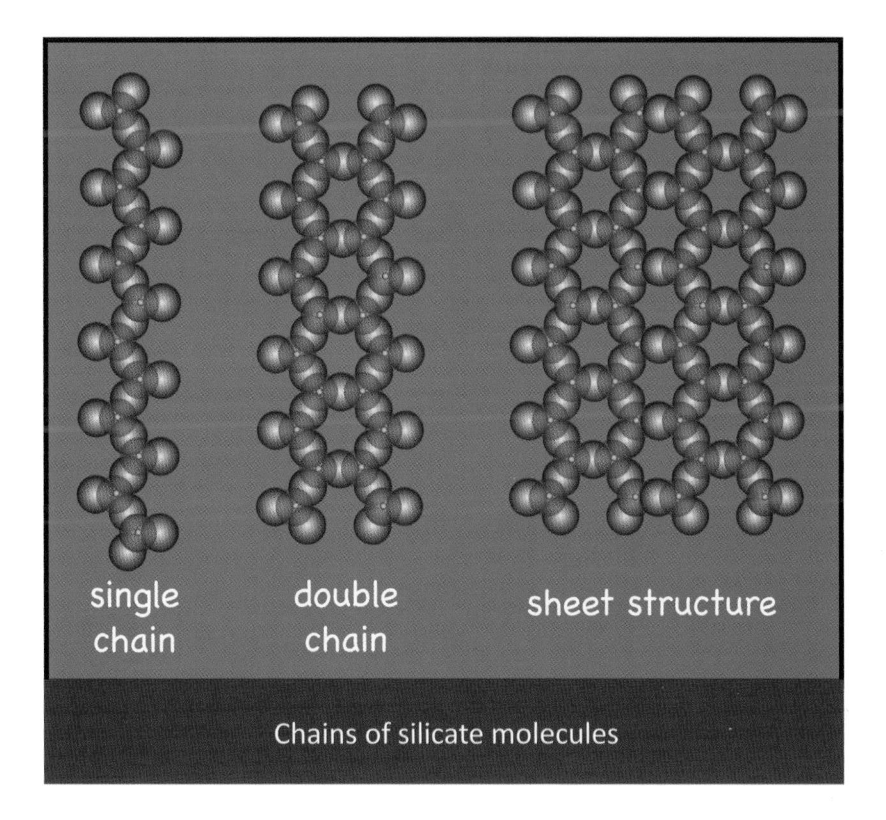

Chains of silicate molecules

Two common silicate minerals are quartz and feldspar. Quartz is made of silicon and oxygen and is found in many rocks and soils. Feldspar contains not only silicon and oxygen but also aluminum and either potassium, calcium, or sodium.

Mica

Mica is another silicate and contains oxygen, silicon, and either potassium, magnesium, iron, or aluminum. Mica is very soft and comes apart in layered sheets.

Not all minerals found in rocks and soils are silicate minerals. Non-silicate minerals found in rocks and soils include calcite, dolomite, halite, and gypsum.

Non-Silicate Minerals
Calcite | Dolomite | Halite

3.3 Rocks

Rocks are aggregates (mixtures) of different minerals glued together by the high heat and pressures found inside the Earth. There are three basic types of rocks called igneous, metamorphic, and sedimentary. Each of these rock types differs from the others based on their chemical makeup, how they were formed, and their texture.

Granite

Igneous rock makes up most of Earth's crust. Igneous rocks are formed from the molten (melted) magma deep within the Earth's core. Magma is made of rocks that have melted due to high temperatures within the Earth. Magma is mostly made of the chemical compound silicate which, as we now know, forms the silicate minerals.

Igneous rocks come in a variety of textures. Some igneous rocks are very coarse and contain large grains of silicates. Other igneous rocks are fine or even glassy, with small grains of silicates.

Granite is probably the best known igneous rock. Because granite has a natural beauty when it is cut and polished, it has been used as kitchen countertops in homes, for monuments, and as building stones. Granite is composed mainly of quartz and feldspar. The quartz particles are often clear and round in shape, and the feldspar is typically rectangular and pink. This combination of shapes and colors gives granite its beautiful texture.

Sedimentary rocks are made from particles of igneous and other rocks that have been weathered (worn away) by physical and chemical interactions. The word sediment comes from the Latin word *sedimentum* which means "to settle," so sedimentary rocks are those rocks formed from materials that are deposited by wind or water and then "settle." Sedimentary rocks are often layered.

There are two main types of sedimentary rocks: detrital sedimentary rocks and chemical sedimentary rocks. Detrital comes from the Latin word *detritus* which means "to wear away," so detrital sedimentary rocks are those rocks that have formed from layers of debris that has weathered, or been worn away, from other rocks. Detrital sedimentary rocks may also contain other debris such as shells and plant matter. These layers of debris turn into rock as they are subjected to pressure from the weight of layers that form above them. Detrital sedimentary rocks include shales and sandstones.

Chemical sedimentary rock formations in a cave

Chemical sedimentary rocks are formed when minerals precipitate from the solution they are dissolved in. This means that the minerals separate out of the solution, most often from evaporation of the water. Limestone is a type of chemical sedimentary rock.

Metamorphic rocks are formed when one rock type transforms into another because of heat, pressure, or chemical reactions. *Meta* is a Greek prefix meaning "along, with, or between," and morph comes from the Greek word *morphe* and means "shape." Metamorphosis means "between shapes" and refers to the transformation or change of one thing into another. Metamorphic rocks are made from preexisting igneous rocks, sedimentary rocks, and even other metamorphic rocks.

Metamorphosis of rocks occurs in different ways depending on the environment they're in. Below the Earth's surface, rocks may be transformed by heat when they come in contact

Metamorphic Rocks

Schist

Marble

with magma. Rocks may also be exposed to hot iron-rich water or acids and bases, causing chemical changes. Mountains can form when rocks encounter extreme pressures.

Metamorphic rocks have different types of texture. Some metamorphic rocks are foliated, or layered. The word foliated comes from the Latin word *foliare* which means "leaf." As a verb, the word foliate means to divide into layers, like the "leaves of a book." The amount of foliation refers to how layered the texture of the metamorphic rock appears. Some rocks are highly foliated, and some are not foliated at all.

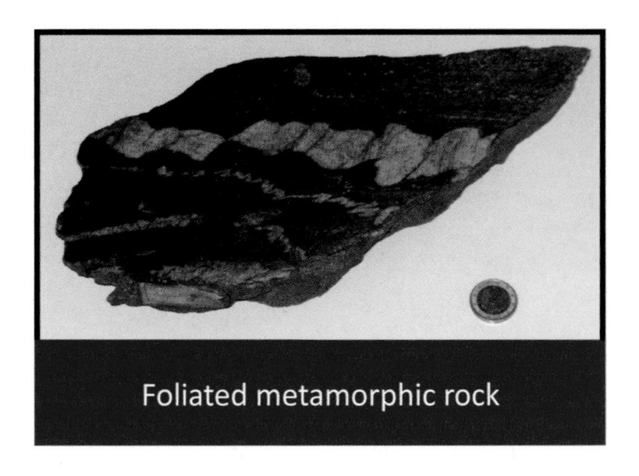

Foliated metamorphic rock

Slate is a common metamorphic rock. The sedimentary rocks shale, mudstone, and siltstone, when exposed to high pressures, become slate. Because slate is a layered rock, it can be easily split into flat slabs. When slate splits, the flat slabs created can be used for patios and walkways and as a roofing material.

Marble statue of Socrates
Louvre Museum, Paris, France

Marble is also a metamorphic rock. Marble is created when limestone is exposed to heat and pressure. Pure marble has a beautiful, naturally white color and has been used for monuments and by artists for sculptures.

Schist is a common metamorphic rock. Schist contains minerals such as mica and quartz. Schist is characterized as foliated because the individual minerals can split easily into flakes.

In summary, the three basic types of rock are igneous, sedimentary, and metamorphic. These rock types differ based on how they were formed and their texture.

3.4 Testing Rocks and Minerals

<table>
<tr><th colspan="5">Three Basic Types of Rock</th></tr>
<tr><th>Rock Type</th><th>Igneous</th><th colspan="2">Sedimentary</th><th>Metamorphic</th></tr>
<tr><td>Formation</td><td>Formed from molten magma; made mostly of silicon and oxygen</td><td colspan="2">Formed from chemical and physical weathering processes</td><td>Formed from the transformation of one rock type to another</td></tr>
<tr><td rowspan="2">Textures</td><td rowspan="2">Fine
Coarse
Large crystals
Glassy</td><td>Detrital</td><td>Chemical</td><td rowspan="2">Foliated
Weakly foliated
Non-foliated</td></tr>
<tr><td colspan="2">Coarse
Medium
Fine</td></tr>
</table>

A rock and mineral test kit can be used to help determine the type of mineral in a sample. A typical rock and mineral test kit contains a Mohs scale of mineral hardness, a penny, a nail, a streak plate, a dropper bottle, and a glass plate.

One identifying feature of minerals is how hard they are. Since a harder mineral will scratch a softer mineral, a scratch test can be used. In 1812 a German geologist named Friedrich Mohs developed the Mohs scale of mineral hardness which is a chart of the relative hardness of ten minerals (see next page). A rock or mineral sample can be tested by rubbing it with a known mineral to see if the known mineral will leave a scratch on the sample. The Mohs scale also assigns a hardness to objects such as a fingernail, a copper penny, a steel nail, window glass, and a steel file. Determining whether one of these objects will scratch the sample or whether the sample will scratch the object can help in identifying the mineral.

Mohs Scale of Mineral Hardness		
Hardness	**Mineral**	**Scratch Test**
1	Talc	can scratch with fingernail
2	Gypsum	can scratch with fingernail
3	Calcite	can scratch with copper penny
4	Fluorite	can scratch with steel nail and glass
5	Apatite	can scratch with steel nail and glass
6	Orthoclase	can scratch with steel file
7	Quartz	may scratch with steel file
8	Topaz	will scratch quartz
9	Corundum	will scratch topaz
10	Diamond	will scratch corundum

Mohs Scale: Hardness of Objects Used for Testing	
Hard-ness	**Test Object**
2.5	fingernail
3	copper penny
5.5	steel nail, window glass
6.5	streak plate
7	Steel file

Hematite rubbed on a streak plate
Hematite photo credit: NASA/JPL

A streak plate is an unglazed porcelain (white ceramic) tile used to show the actual color of the mineral sample. When an edge of the sample is rubbed across the streak plate, it leaves a colored streak. The colored streak left on the plate is actually the color of the powdered mineral which may be different from the appearance of the mineral sample as a whole. One example is hematite, an iron-containing mineral, which can have a gray color as a rock but when rubbed on a steak plate will leave a red mark. About 20% of minerals will leave a colored streak with about 80% leaving a white or clear streak.

A rock sample can also be tested to see if it will have a chemical reaction with acid. The acids most often used are dilute hydrochloric acid or acetic acid (vinegar). The dropper bottle in the test kit is used to apply the acid to the sample. For instance, when an acid is applied to rocks containing the mineral calcite, the chemical reaction can be observed by the formation of bubbles of carbon dioxide gas.

A rock and mineral test kit may also contain a sample of each of the ten minerals that appear on the Mohs scale of mineral hardness. These can be used for scratch testing unknown samples.

3.5 Soils

Soils are composed of different types of materials including weathered rocks and minerals and the remains of plants and animals. The source of the weathered matter from which soils develop is called the parent material.

The parent material affects the type of soil that is created. Some soils will support plant growth and some will not, according to the parent material from which they were formed. For example, soils developed from granite or sandstone do not support plant growth as well as do soils developed from limestone. Limestone is made of calcium carbonate, and calcium creates a nutrient-rich soil for plants to grow in.

Other factors that affect soils are climate, time, and topography. Topo comes from the Greek word *topos* and means "place." Graphy comes from the Greek word *graphe* which means "to write," so topography means the "description of the place." Topography describes the places where soils are developed (sloped mountain ranges, riverbeds, swamp beds, etc.), which also determine the type of soil formed.

In order to organize the different types of soils, scientists have created a soil classification system called soil taxonomy. The soil taxonomy system is similar in some ways to the taxonomy system used for living things. Both taxonomy systems attempt to organize a broad spectrum of characteristics into meaningful categories. Soil taxonomy categorizes the different types of soils found on Earth according to different soil properties, such as soils that provide nutrients for plant growth or soils that are made from volcanoes or clay.

The twelve main categories for soils are called soil orders. The names of soil orders are derived from Latin and Greek words, just like the names for biological organisms. For example, the order for dry soils is Aridisols which comes from the Latin words *aridus* which means "dry" and *solum* which means "soil." The order for prairie soils is called Mollisols which comes from the Latin word *molere* which means "to grind or crush." Mollisols are soils obtained from parent materials that have been ground or crushed, and these soils are loose and soft.

World Soil Orders

Soil order	Description	Soil order	Description
Alfisols	high-nutrient soils	**Inceptisols**	young soils
Andisols	volcanic soils	**Mollisols**	prairie soils
Aridisols	desert soils	**Oxisols**	high nutrient soils
Entisols	new soils	**Spodosols**	conifer forest soils
Gelisols	permafrost soils	**Ultisols**	low-nutrient soils
Histosols	organic soils	**Vertisols**	swelling clay soils

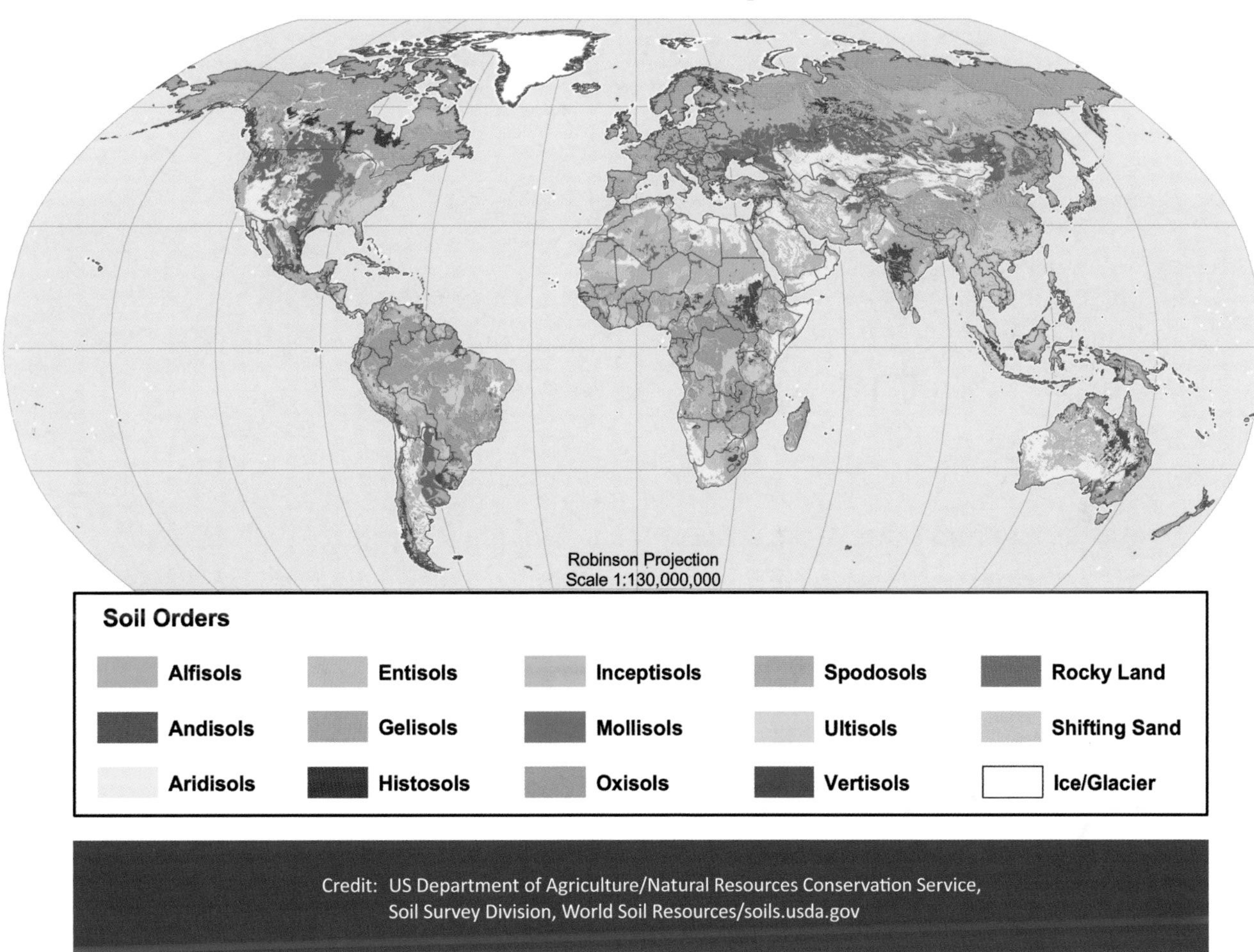

3.6 Summary

- The surface of the Earth is made of rock, minerals, and soils.
- In order for a material to be considered a mineral it must meet the following criteria:

    1) It must occur naturally.
    2) It must be a solid.
    3) It must be inorganic.
    4) It must have an ordered arrangement of atoms.

- The eight elements that make up the majority of rock-forming minerals are: oxygen, silicon, aluminum, iron, calcium, potassium, sodium, and magnesium.

- The three basic types of rock are: igneous, metamorphic, and sedimentary.
- Soil taxonomy is a classification system that sorts different types of soils into categories according to their properties.

3.7 Some Things to Think About

- Go outside and take a walk. Bring a notebook (or your field notebook) and write descriptions of some of the different types of rocks and soil you observe. Also note anything that is surprising or not what you expected.
- Is the air we breathe a mineral? Why or why not?
- Go outside and find some rocks. Inspect them, looking for features such as color, texture, layering, uniformity, etc. Use a magnifying glass if one is available. Review the descriptions of igneous, sedimentary, and metamorphic rocks and guess which type of rock you are looking at. Record your observations.
- Test the rocks you gathered. Try rubbing them against each other. Do any of the rocks scratch other rocks? Do any of them crumble? Can you tell which rock is the hardest? The softest?
- Do you think that farming/agricultural has an effect on the soil? Why or why not?
- Do you think it would help a farmer to know what kinds of soil are on the farm? Why or why not?

Chapter 4 Earth's Layers

4.1 Introduction

We have seen that Earth is made of minerals, rocks, and soils, but how do they fit together to make a planet?

How far down do the rocks and soil go? Are there more rocks beneath the soil, or is there only empty space? What is in the middle of the Earth? Are rocks and soil all that make up the Earth, or is there more to the Earth? What about water, the air, and sky?

4.2 Inside the Earth

Scientific evidence suggests that Earth has layers, much like a golf ball. If you examine a golf ball sliced in half, you can see that the outer layer is different in texture and composition from the inner layers.

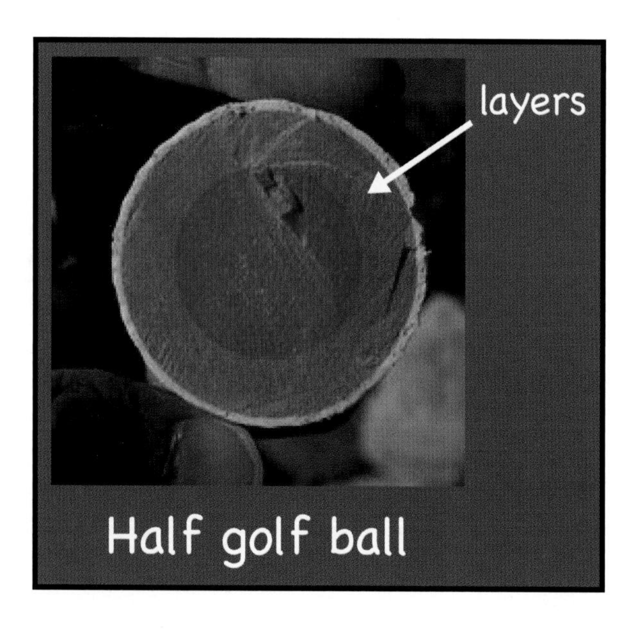

Half golf ball

Similarly, the Earth's layers are different from each other. The layers can be divided based on physical properties. The differences in physical properties include whether or not a layer is solid or liquid, how weak or how strong it is believed to be, and what it is made of. The Earth is considered by many scientists to consist of three distinct main layers—the crust, the mantle, and the core. The mantle and the core are further divided into layers.

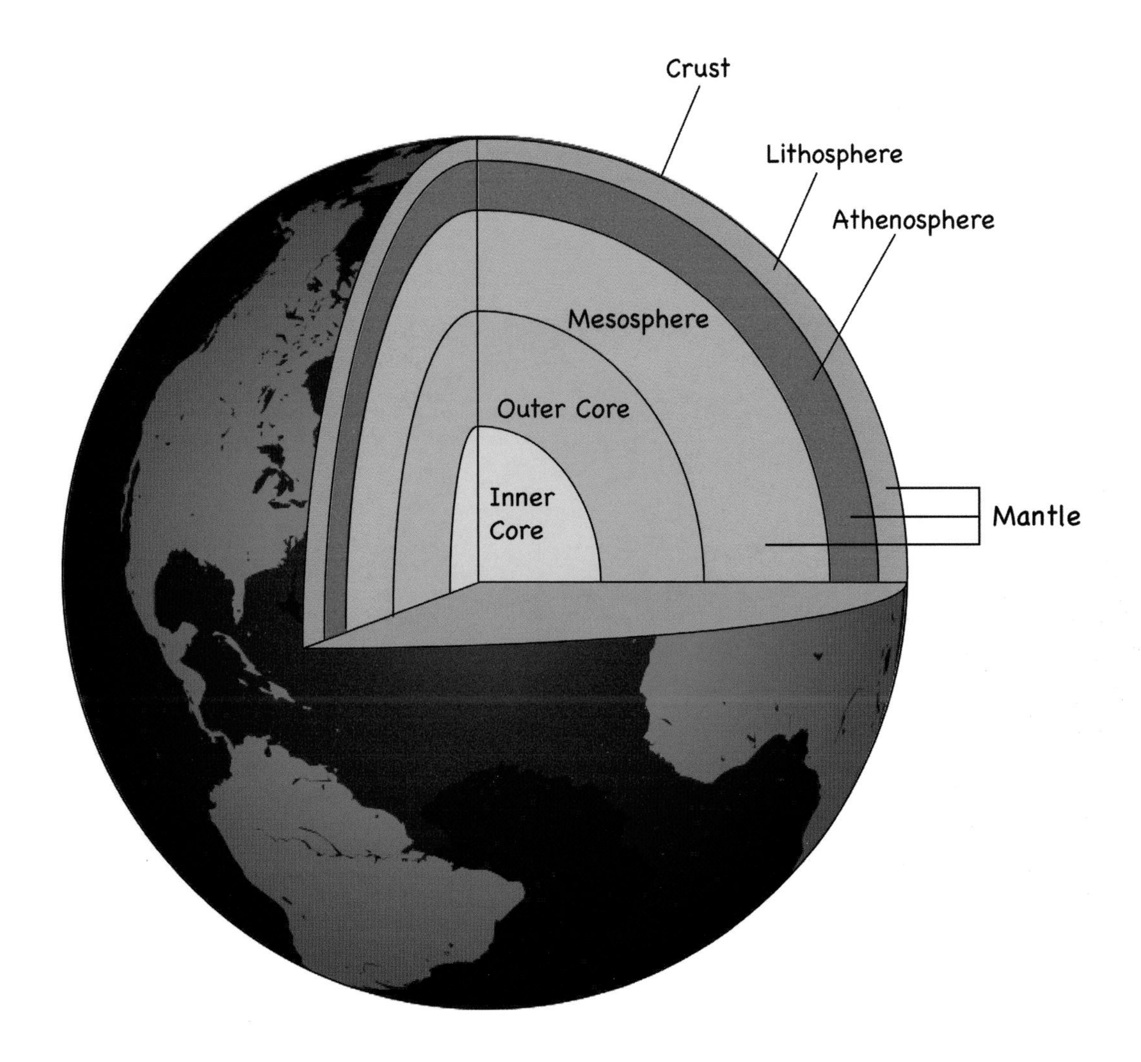

4.3 The Crust

Rocks and soil make up the Earth's surface, which is called the crust. The crust is often referred to as the outer shell of the Earth, just like the white layer surrounding a golf ball is its outer shell. Earth has two different types of crust — the continental crust and the oceanic crust.

The continental crust forms the continents, or land masses, on the surface of the Earth. The continental crust lies above sea level and is home to land-dwelling plants and animals. The continental crust averages about 35-45 kilometers (22-28 miles) in thickness and is composed mainly of granite but also includes other igneous rocks as well as sedimentary and metamorphic rocks.

The oceanic crust is the part of the Earth's surface that lies below the oceans. The oceanic crust is much thinner than the continental crust—averaging about 7 kilometers (5 miles) thick—and is composed mainly of basaltic rock, an igneous rock formed from cooled magma. The oceanic crust has an upper layer of sediment that has been deposited on the basaltic rock layer.

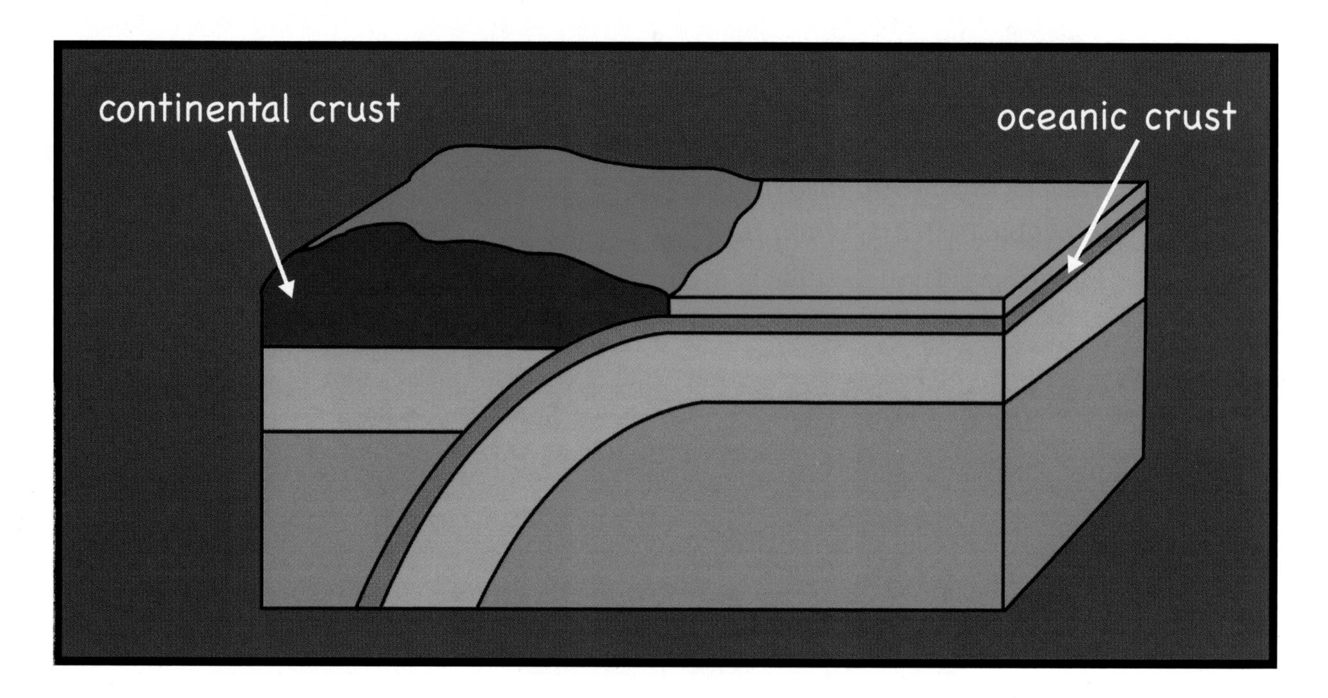

4.4 The Mantle

The mantle is the layer of Earth just below the crust. The mantle is further divided into three layers which are called the lithosphere, asthenosphere, and mesosphere.

It's a little hard to know exactly what lies below the Earth's crust because no one has been able to drill deep enough to find out. The deepest land-based hole that has been drilled is on the Kola Peninsula in Russia and reaches a depth of about 12 kilometers (7 miles). There is also a research vessel named Chikyu that is designed to drill holes in the ocean floor where the crust is thinner, and the hope is that one day it can drill below the oceanic crust.

So far no one has been able to drill deep enough to directly sample Earth's inner layers. What we know of the material below the Earth's crust is based on data taken from a variety of observations including volcanoes, earthquakes, and seismic readings. By detecting vibrations caused by earthquakes, seismic readings record movements of materials within the Earth.

4.5 The Lithosphere

Just below the crust is the lithosphere which is the upper layer of the mantle. The prefix litho comes from the Greek word *lithos* and means "stone." Lithosphere literally means "stone sphere." The crust and lithosphere are sometimes considered together as one layer.
The lithosphere is believed to be a relatively cool, rigid shell about 100 kilometers (62 miles) thick. Scientists think that the lithosphere is divided into separate regions or plates. The crustal land masses sit on top of these plates, allowing the land masses to move.

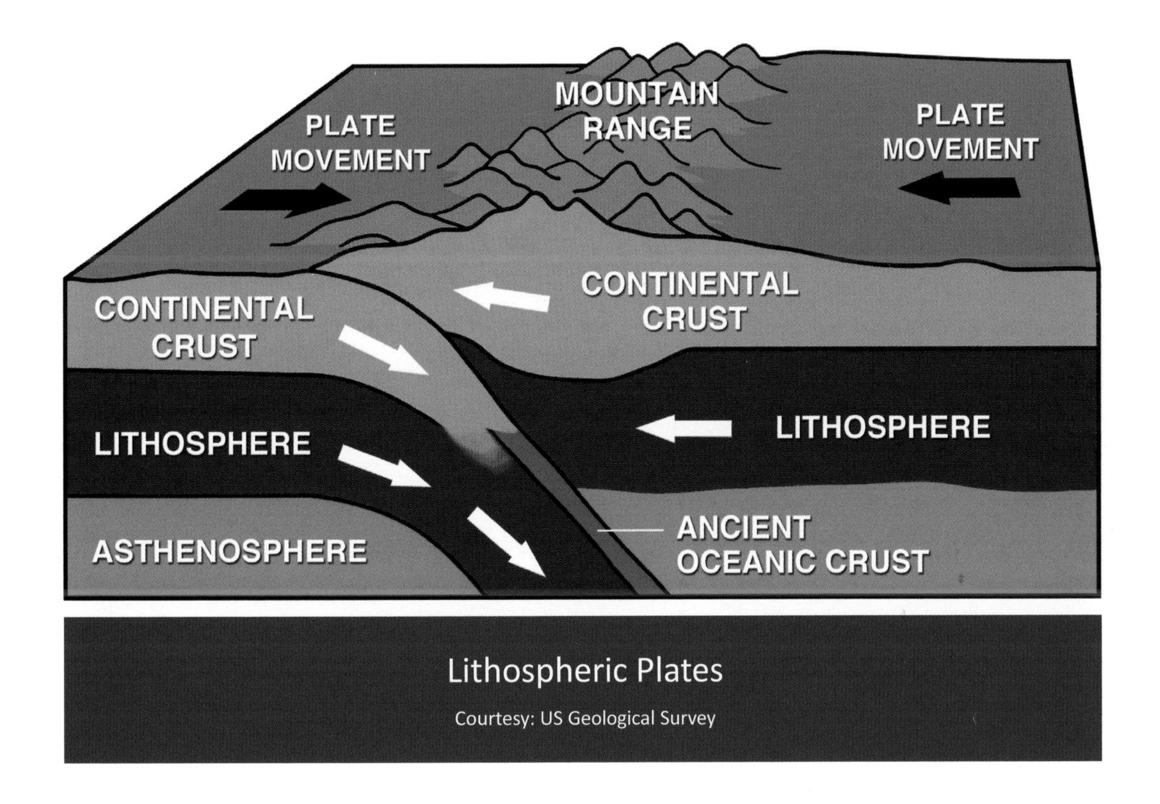

Lithospheric Plates

Courtesy: US Geological Survey

Although the lithosphere is believed to be made of different kinds of chemical materials, there is evidence to suggest that this layer is rigid and inflexible. For example, when different plates scrape against each other, the plates do not deform, or change shape, significantly. This would indicate that the plates are rigid. Also, volcanic activity tends to cluster at the edges of the plates with little or no activity occurring in the center. Again, this supports the theory that the plates are rigid enough to prevent the molten material below the lithosphere and crust from penetrating those layers and coming to the surface.

4.6 The Asthenosphere

The lithosphere sits on a softer layer called the asthenosphere. The prefix astheno- comes from the Greek word *asthenes* which means "weak." Asthenosphere literally means "weak sphere."

The rock at the top part of the asthenosphere is under extreme pressure due to the weight of the lithosphere and crust above it. Scientists believe that this extreme pressure creates high heat and causes the minerals in the rock to melt. Because of the high heat and extreme pressure, the asthenosphere is believed to have a putty-like texture, allowing it to move around.

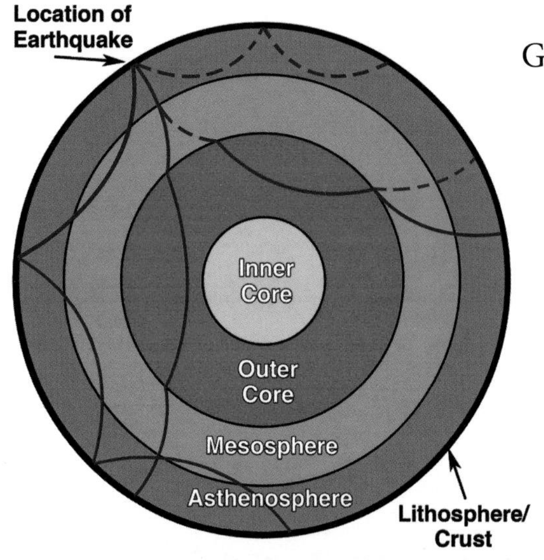

Seismic waves from an earthquake travel through Earth

Derived from US Geological Survey illustration

Geologists are unable to collect much information about the asthenosphere because, like the lithosphere, it is too deep to sample directly. However, the way seismic waves move through the various layers can be studied, and changes in the speed of seismic waves as they move through the Earth can be detected. This change in speed indicates a transition that suggests a boundary between layers. By observing how seismic waves travel and change their speed, geologists can get some idea of the type of material in the asthenosphere.

Based on seismic wave data, scientists believe that the asthenosphere may extend below the surface of the lithosphere as far as 700 kilometers (435 miles) and that the rigid lithospheric plates are able to move because they sit on top of the plastic, putty-like asthenosphere. It is also thought that currents in the Earth's inner layers cause these plates to move.

If you watch a pot of water come to a boil, you will observe the water moving and swirling just before it begins to boil. This moving and swirling occurs because of convection. Convection is the transfer of heat by the movement of molecules. When heat is applied to the water, energy is being added. However, water rarely heats uniformly, so some water molecules will be moving fast (have more energy) and other water molecules will be moving more slowly (have less energy). When this happens, energy is transferred from molecule to molecule as they bump into each other, causing the water to swirl. In a similar manner, convection occurs within the Earth's layers.

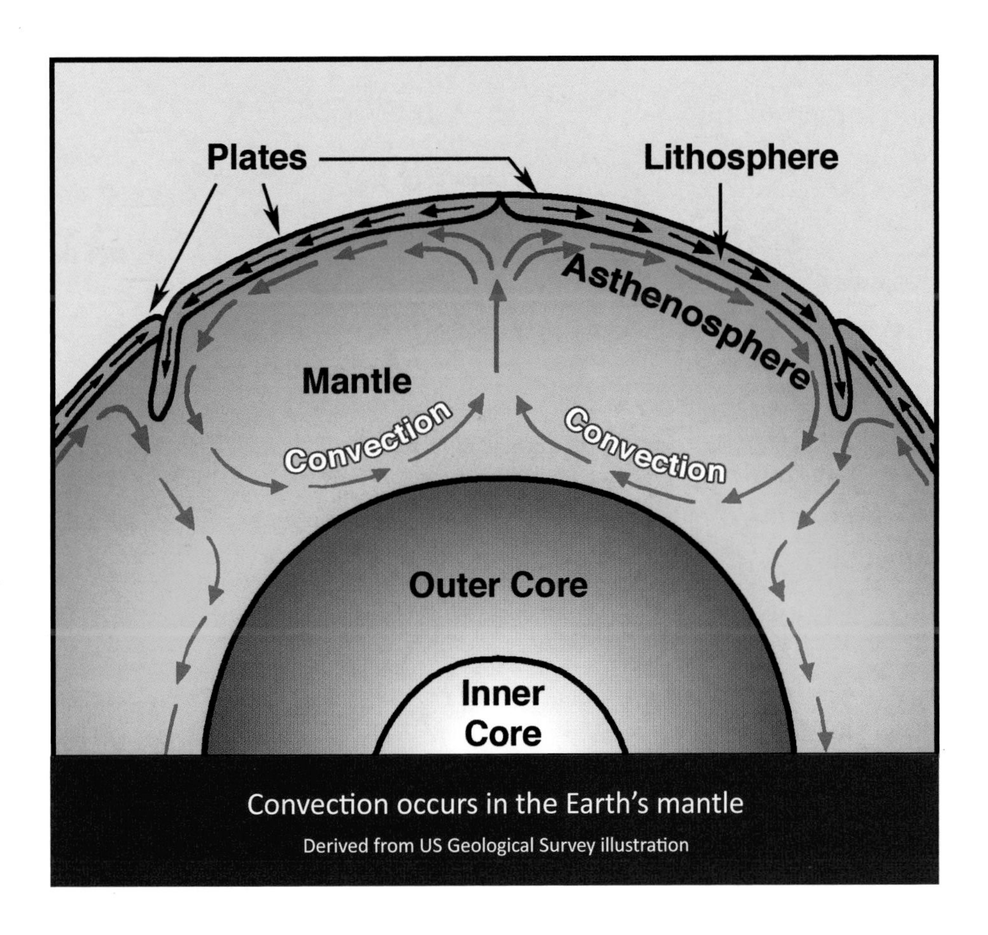

Convection occurs in the Earth's mantle
Derived from US Geological Survey illustration

4.7 The Mesosphere

Below the asthenosphere lies the mesosphere. Meso comes from the Greek word *mesos* and means "middle" or "between." The mesosphere is the layer between the core and the outer layers of the Earth.

The mesosphere is believed to extend to a depth of 2900 kilometers (1800 miles). Many scientists believe that even though the temperatures are higher in the mesosphere than the asthenosphere, the rocks in the mesosphere are more solid due to higher pressures.

4.8 The Core

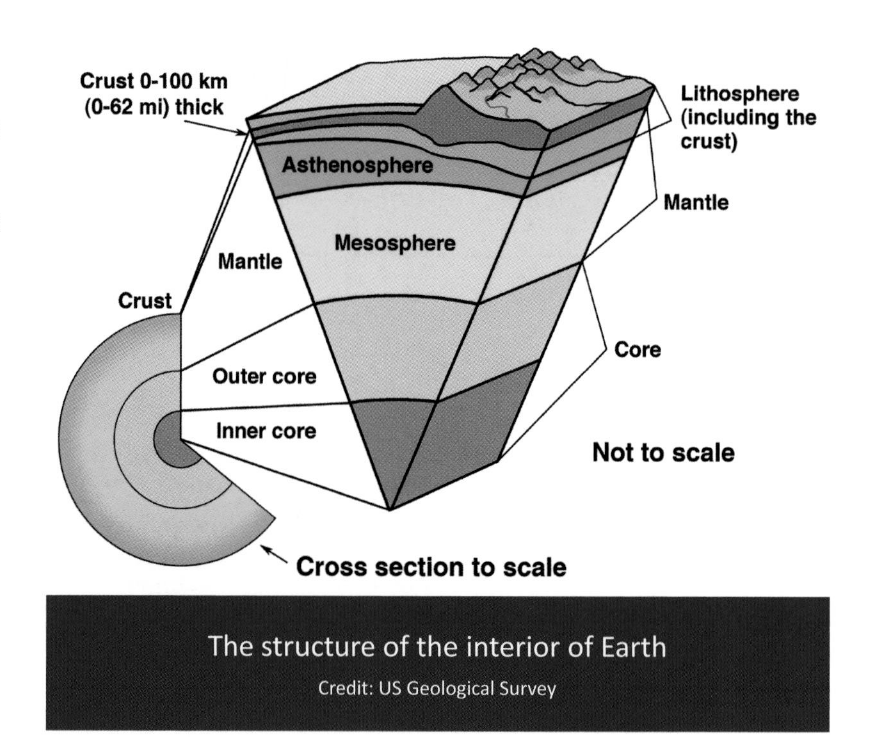

The structure of the interior of Earth
Credit: US Geological Survey

The core is at the center of the Earth and is divided into two layers—the outer core and the inner core. These two layers are believed to be composed of the elements iron and nickel.

The outer core seems to behave like a liquid layer and is thought to be about 2270 kilometers (1410 miles) thick. The inner core behaves more like a solid layer with a radius of about 1200 kilometers (745 miles). The inner and outer core layers together account for about one-sixth of Earth's volume and nearly one-third of its total mass. The density of the core is about 14 times the density of water. (Density is the amount of mass contained in a certain amount of three-dimensional space and is expressed mathematically as mass divided by volume.) This means that there is a large amount of matter packed very tightly into the core layers, making the core very dense.

It is thought that the temperature of the core can exceed 6700° C (12,092° F), and its pressure is millions of times greater than the pressure at the Earth's surface.

Convection occurs in the Earth's outer core. The hot liquid iron and nickel move and swirl because of convection. Many scientists think that heat convection happening in the outer core is responsible for many of the dynamic processes that occur on Earth's surface, such as earthquakes and volcanoes. Earth's core is also thought to create Earth's magnetic field.

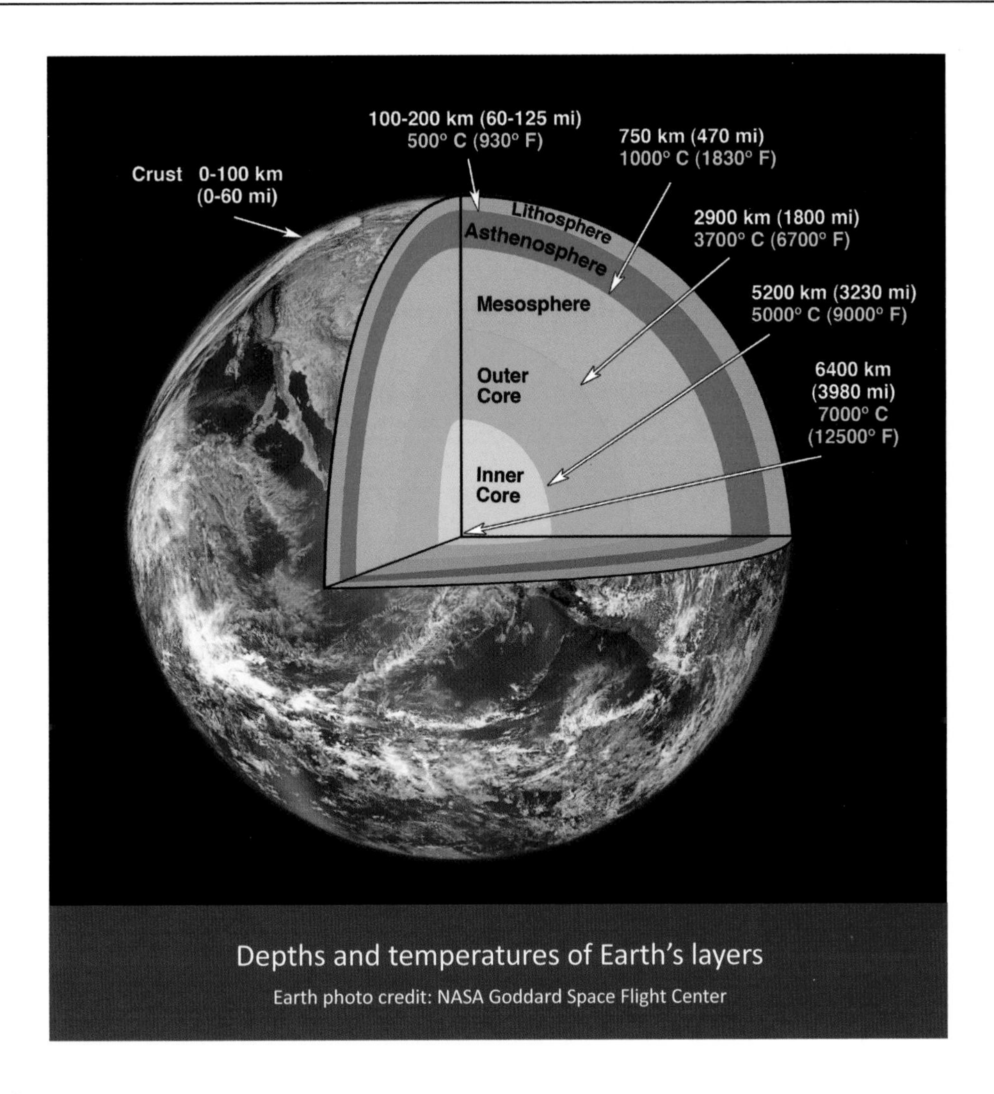

Depths and temperatures of Earth's layers

Earth photo credit: NASA Goddard Space Flight Center

4.9 Summary

- The Earth is thought to be made up of three main layers—the crust, the mantle, and the core.
- We live on the outer layer called the crust.
- The mantle is just below the crust and is made up of three layers — the lithosphere, the asthenosphere, and the mesosphere.
- The core is in the center of the Earth and has two layers which are called the outer core and the inner core.
- Geologists study volcanoes, earthquakes, and seismic readings to help them understand what lies below the Earth's surface.

4.10 Some Things to Think About

- Imagine that you could dig a big hole all the way through the Earth and out the other side. What do you think you would find as you traveled through the hole?
- What do you think causes the interior of Earth to have different layers?
- If you wanted to drill a really deep hole to find out what is below the crust, would you drill through the continental crust or the oceanic crust? Why?

  Do you think it is possible to drill a hole that is deep enough to go all the way through the crust?
- How do you think volcanoes can help geologists find out what the mantle is like?
- What do you think makes volcanoes more likely to occur along the edges of tectonic plates than in the center of the plates?
- If the asthenosphere were solid, do you think earthquakes and volcanoes could occur? Why or why not?
- What do you think creates the pressure in the mesosphere?
- Do you think all the layers of Earth work together and interact with each other, or is each layer on its own with no relationship to the others? Why?

Chapter 5 Earth's Dynamics

5.1 Introduction

We know from experience that the Earth is dynamic and is constantly changing. Earthquakes, volcanoes, tsunamis, and other natural phenomena remind us that there is movement in the crust of the Earth and what is below it.

How do these changes occur? What causes earthquakes? During a volcano where does the hot molten rock come from? What happens during a tsunami? How do mountain ridges form? All of these questions address Earth's dynamics.

Geologists study the eruption of Mount St. Helens, Washington
Photo Credit: US Geological Survey

5.2 Plate Tectonics

Explaining what makes earthquakes, mountain ridges, and volcanoes occur can be a daunting task. But, by analyzing data and observing changes on the Earth, scientists have developed the theory of plate tectonics to help them understand these events. In general, the theory of plate tectonics explains the movement of the lithosphere, and consequently, the movement of Earth's land masses, the formation of mountains, and earthquake and volcanic activity.

Scientists have found evidence to suggest that the lithosphere (the layer of the Earth just below the crust) is broken up into rigid plates. The crust sits on top of these plates, which are "floating" on top of the softer, putty-like asthenosphere. These plates move very slowly as the material in the asthenosphere circulates due to convection.

According to plate tectonics, there are seven primary plates and several smaller secondary and tertiary plates that fit together like pieces of a jigsaw puzzle. Heat convection from the core and lower mantle causes the plates to move very slowly at a rate of about 2 inches per year.

One line of evidence that supports the theory of plate tectonics is the observation that earthquakes and volcanoes congregate along certain regions of the globe. These regions are believed to be boundaries between the plates.

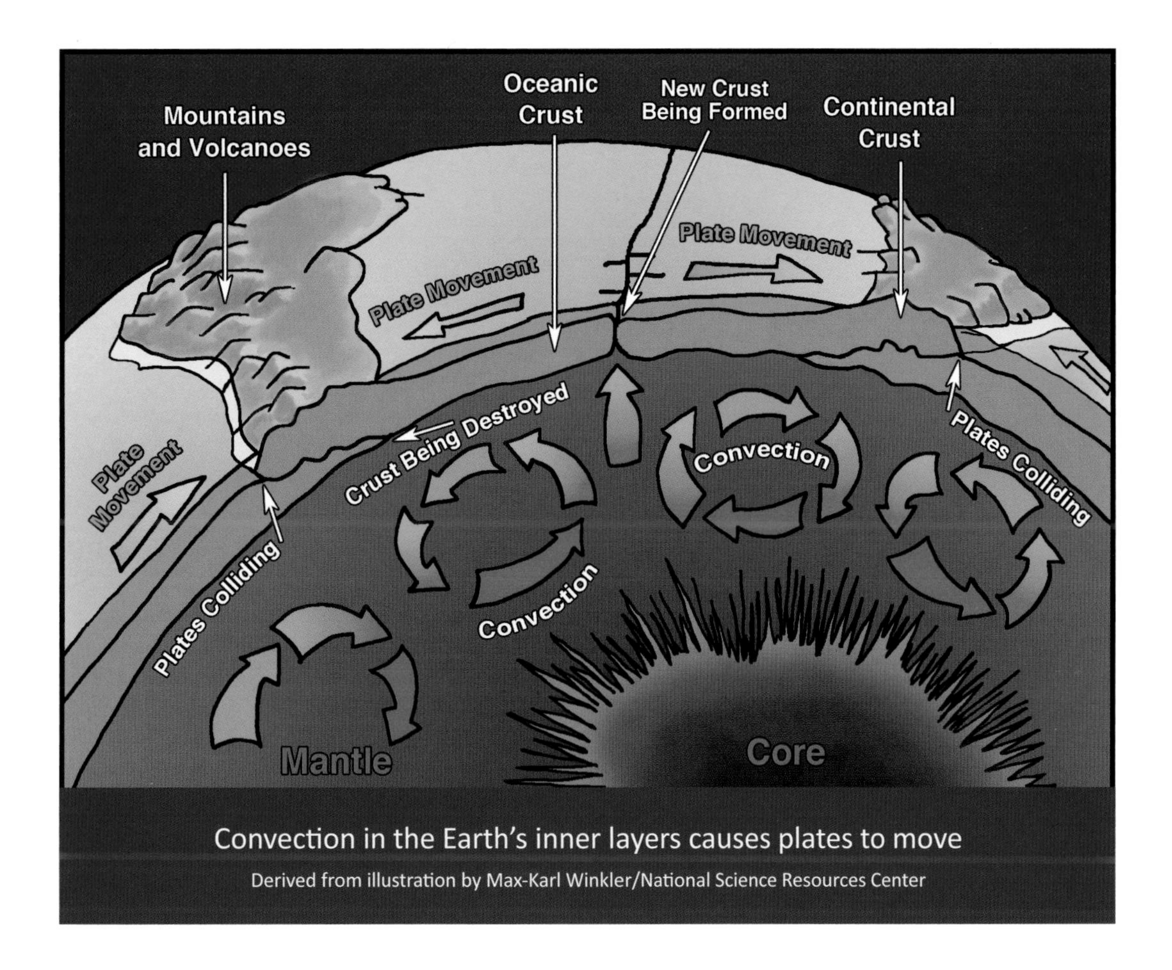

Convection in the Earth's inner layers causes plates to move

Derived from illustration by Max-Karl Winkler/National Science Resources Center

Curiously, much of the data supporting the theory of plate tectonics were collected shortly after World War II and during the Cold War. Countries, being suspicious of each other, developed worldwide seismograph networks to monitor underground nuclear testing. As a result, we now have a map of the areas of the globe that have the most earthquake activity.

Countries spent millions of dollars mapping the ocean floor to find safe ways to navigate and to discover ways to detect submarines. This research provided us with knowledge of the areas where volcanic activity dominates and where underwater mountain ridges associated with volcanic activity are located. Much of the data that was collected supports the theory of plate tectonics and the idea that Earth's land masses move on top of rigid lithospheric plates floating on the putty-like asthenosphere.

It should be noted that although this theory is useful in explaining many large-scale geological processes, plate tectonics is a theory that is still developing, and not everything is understood. As new data become available, the current theory will likely change and be refined.

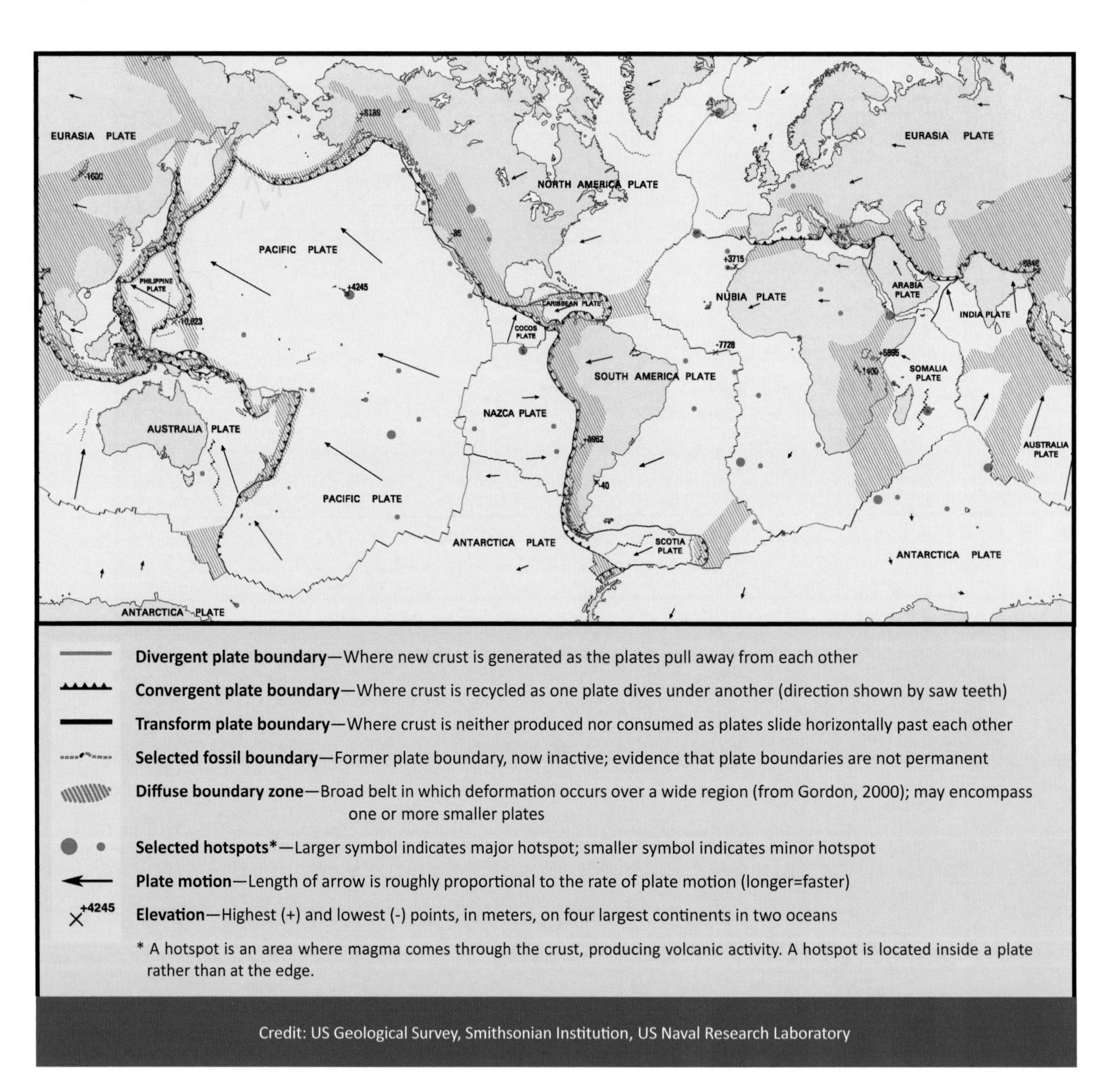

Credit: US Geological Survey, Smithsonian Institution, US Naval Research Laboratory

5.3 Mountains

There are many different types of mountains. There are steep, jagged mountains and smooth, rounded mountains. There are mountains covered with lots of trees and mountains with rock alone and no vegetation.

How do mountains form? There are actually several different ways in which mountains are created. Mountains can form when land masses are pushed together and folded onto each other as a result of the movement of plates. These are called folded mountains. An anticline results when the land is folded in an arch shape. When land is folded in a trough shape, it is called a syncline.

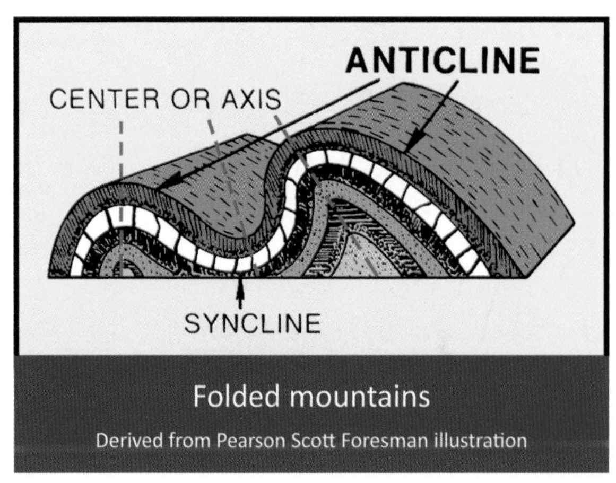

Folded mountains
Derived from Pearson Scott Foresman illustration

Fault-block mountains can form as a result of big blocks of land moving up and down. When plates slide up and down with respect to each other, a huge block of land is pushed upwards and another huge block of land is pushed downwards. The pushing up and down of these blocks can create fault-block mountains and also can cause earthquakes.

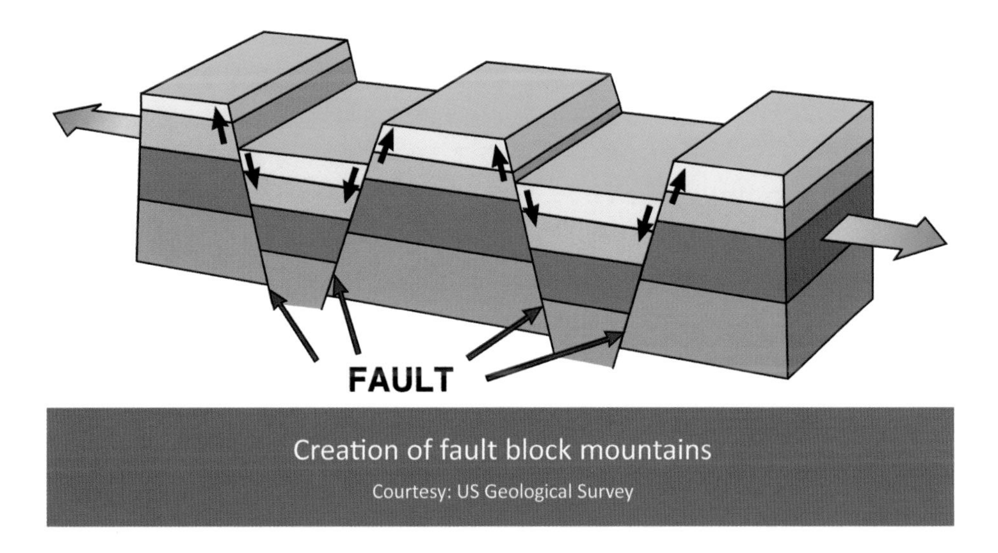

Creation of fault block mountains
Courtesy: US Geological Survey

Mountains can also form when molten rock, or magma, pushes up underneath the crust but doesn't break through the surface. Instead, the force of the magma creates a large bump on the Earth's surface. The magma cools, leaving a smooth, rounded landform called a dome mountain.

Lava fountain and lava flows
Kilauea Volcano, Hawaii Volcanoes National Park
Photo Credit: US Geological Survey/by J. D. Griggs

5.4 Volcanoes

Volcanoes are fascinating because we get to see the molten rock that comes from deep inside the Earth! We also get to witness the extreme pressures that force the molten material through the surface of the Earth. A volcano forms when a fracture, or break, in the upper mantle allows magma from below to be pushed through the surface of the Earth. When magma comes to the surface of the Earth, it is called lava.

Volcanic material, or magma, is mainly basaltic rock and is believed to come from the asthenosphere, the soft putty-like layer below the lithosphere. Some volcanic eruptions are "gentle," like the slow moving lava flows in Hawaii, and others are more violent, such as the eruption of Mount St. Helens in the state of Washington in 1980. The degree to which a volcano will erupt gently or violently depends on different properties of the magma, such as its viscosity (thickness or thinness), how much gas it contains, the minerals it is made of, and its temperature.

Lava flow, Hawaii
Photo Credit: US Geological Survey/
Hawaiian Volcano Observatory

There are several different types of volcanoes. A cinder cone volcano is typically a small volcano that forms around a volcanic vent — the opening in the ground that the lava comes through. Cinder cones are formed as lava erupts from the vent in a fountain-like manner and quickly cools into basaltic fragments of rock called cinders.

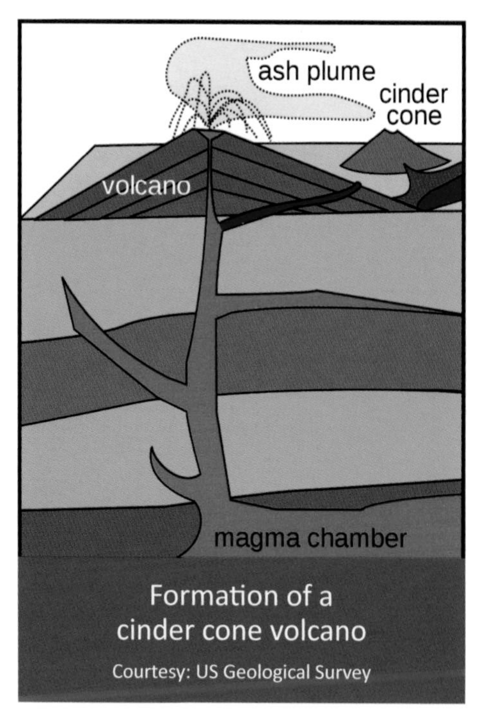

Formation of a cinder cone volcano
Courtesy: US Geological Survey

Cinder cone near Veyo, Utah
Photo Credit: Mark A. Wilson, The College of Wooster

Generally, cinder cones are less than 300 meters high (1,000 ft). The slope of the sides is very steep, and lava often oozes out from the base of the cone during the later part of the eruption.

A shield volcano is built up of layers of low viscosity (thin) lava that can flow over a long distance, resulting in the distinctively long, gentle, sloping sides seen in this type of volcano.

Shield volcano: Mauna Loa Volcano, Hawaii
Photo Credit: Gordon Joly, CC BY SA 3.0

The eruptions that cause shield volcanoes to form are nonexplosive, and shield volcanoes can be extremely tall. The Mauna Loa Volcano in Hawaii is 4,169 meters (13,677 ft) above sea level and 8,534 meters (28,000 ft) above the ocean floor!

Composite volcano, Mount St. Helens, Washington
Photo Credit: US Geological Survey/by Donald A. Swanson

A composite volcano (also called a stratovolcano) has alternating layers of lava flow and lava flow rubble, such as volcanic ash and cinders. Composite volcanoes can be more than 3,000 meters (9,843 ft) high. Their sides are generally steep, and these volcanoes tend to erupt explosively.

5.5 Earthquakes

Have you ever been in a place where suddenly the ground began to move, the surface of the ground split, and buildings began to rumble and rock? If you have, you might have been in an earthquake.

San Francisco Earthquake, 1906
Photo Credit: H. D. Chadwick, NARA National Archives and Records

Earthquakes can be terrifying and many people have experienced their devastating effects. One of the most famous earthquakes to happen in the United States was on April 18, 1906 in San Francisco. In 1906 scientists didn't yet know why earthquakes happen.

When this big earthquake hit, large sections of land became displaced, or moved from their former positions. Fences were moved, homes were ripped apart, and land permanently shifted. This earthquake gave scientists some very important clues to help them understand why earthquakes occur.

Using the theory of plate tectonics, we now suspect that earthquakes occur as a result of neighboring plates colliding, pushing on, or sliding against each other.

These plate movements create force, pressure, and stress along the edges of the plates and within them. This causes fractures, or cracks, in the rocks of the Earth's crust. These fractures are called faults. The term fault line is used to describe the area where the two sides of a fracture meet.

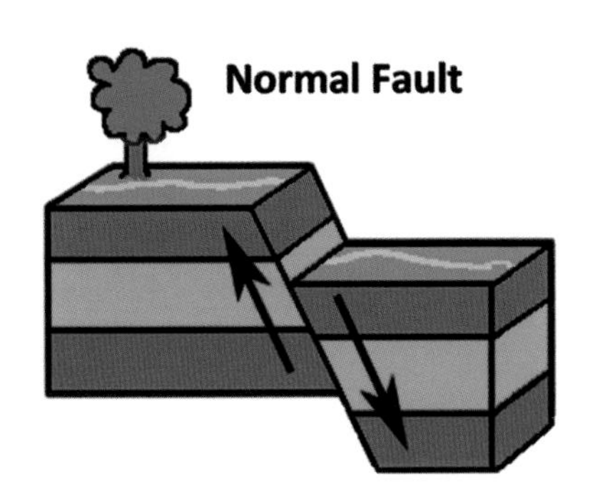

Observations show that earthquakes are caused by movement along the faults in the Earth's crust. Since the faults have rough rather than smooth surfaces, the two sides of a fault get stuck together instead of sliding past each other. As the tectonic plates move, stress (a force resulting from pressure or tension) is created along a fault line. This happens because the two sides of a fault are being pushed on but don't move because they are stuck together. As the plates continue to move, stress builds up along the fault line. When the stress, or stored energy, in the rocks gets too big for the sides of the fault to remain stuck together, the stored energy is suddenly released and the rocks lurch, causing the Earth to shake. The strength of earthquakes can vary from mild to violent depending on the amount of stress that was built up and then released.

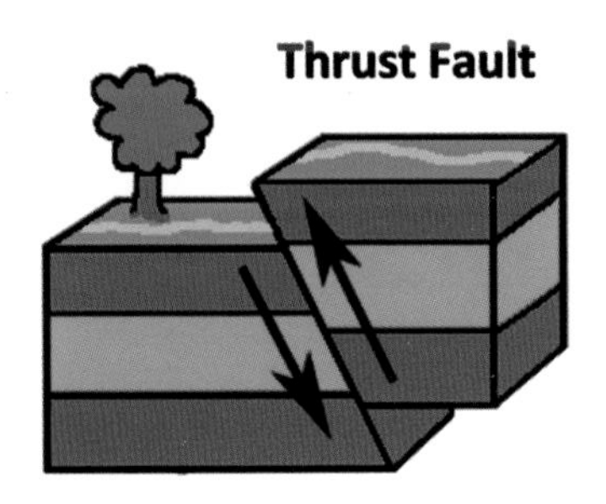

Millions of fault lines crisscross the Earth's crust. Some of these fault lines coincide with tectonic plate boundaries, but many fault lines are only a few kilometers (miles) long and are not located at plate boundaries. Most faults are inactive and don't change very much over time, thus causing little or no earthquake activity. However, some faults are quite active, with earthquakes occurring frequently. This is especially true of faults occurring along tectonic plate boundaries.

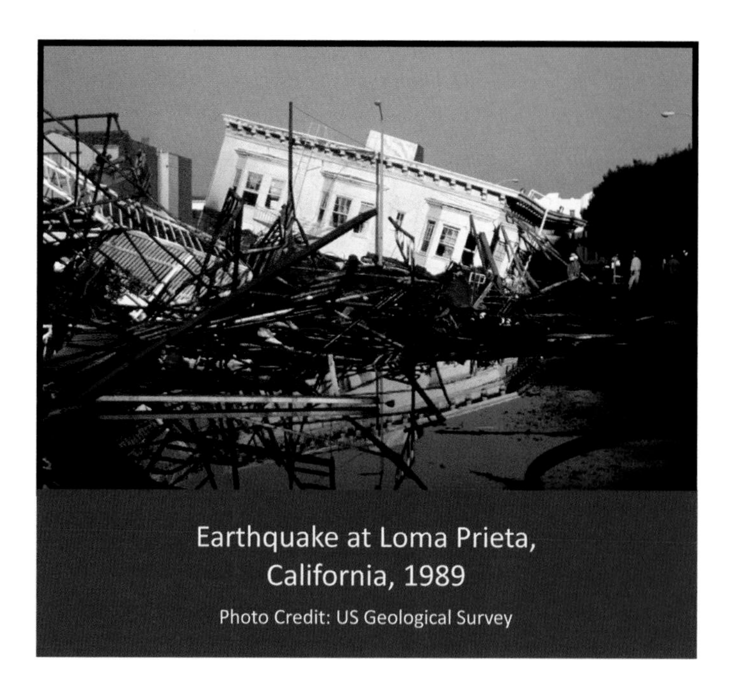

Earthquake at Loma Prieta, California, 1989

Photo Credit: US Geological Survey

Volcanic activity can also cause earthquakes, and large earthquakes can lead to increased volcanic activity.

5.6 Summary

- The Earth is dynamic and constantly changing.
- Plate tectonics is a theory that helps explain Earth's dynamics.
- Mountains can form when land is pushed together (folded mountains), as the result of earthquakes (fault-block mountains), or by pressure created from molten magma (dome mountains).
- Volcanoes occur when magma is pushed through the Earth's surface. Three types of common volcanoes are cinder cone volcanoes, shield volcanoes, and composite volcanoes (stratovolcanoes).
- Earthquakes occur along fault lines due to stresses within the Earth's crust caused by different areas of land moving differently with respect to each other.

5.7 Some Things to Think About

- Do you think Earth is constantly changing or does it stay the same? Why?
- How do you think the development of the theory of plate tectonics helped explain many large-scale geological processes that were not understood before?
- Explain in your own words how different movements of plates create different types of mountains.
- Explain why volcanoes come in different sizes and shapes.
- In your own words, explain the relationship between tectonic plates and earthquakes.
- How do you think volcanic activity can cause earthquakes and earthquakes can increase volcanic activity?

Chapter 6 Earth's Spheres

6.1 Introduction

As you go about your day, you may not think much about all the different parts that make up the Earth. If you live in a house or an apartment, your days are likely illuminated with artificial light rather than just the Sun. If you live in a city, you probably can't see many stars from your bedroom window at night. If you travel from home by car, you can't feel the ground beneath your feet or smell the rain outside. If you shop at the grocery store, you may buy apples, oranges, and bananas, but you probably don't think about the trees, dirt, bugs, and worms that are needed for making the apples, oranges, and bananas grow. And unless you enjoy using a compass to find your way while hiking, you probably don't think about the magnetic field that surrounds the Earth.

However, if you live on a farm or in the woods or wilderness, you are likely to be much more familiar with the different parts that make up Earth. If you've lived long enough on a farm or in the woods or wilderness, you can probably predict rain by the shape of the clouds and know when the soil is just right for planting and which plants do well when planted in the heat of summer and which do better planted when it's colder. You notice that the length of the day varies with the seasons, and you likely know when to expect the first frost so you can harvest crops before they freeze. You may know when the nearby lake will be full of young fish, tadpoles, and ducklings and when it will be frozen over with a layer of ice thick enough to skate on. You probably also know how the rain affects the growth of plants, how animals live by eating plants, and how both plants and animals change the Earth's soil.

In today's modern world, many of us have lost touch with how water, air, ground, plants, animals, and the magnetic field shape and govern Earth's habitats. These features (air, water, ground, plants and animals, and magnetic field) are all part of an interconnected system of spheres that make life on Earth possible.

6.2 The Spheres of Earth

Earth is a system. A system is a set of different parts that work together as a whole. The different parts that make up Earth surround the planet, and because Earth is almost spherical in shape, these parts that surround Earth are referred to as spheres. Although Earth's spheres are sometimes categorized differently, in this textbook we will describe the Earth as being made of five different spheres: the geosphere, biosphere, hydrosphere, atmosphere, and magnetosphere.

The geosphere makes up the rock part of Earth — the crust, mantle, and core. The crust is the solid outer layer of Earth that is made of rocks and minerals and is the part of the geosphere where plants and animals make their home. The geosphere is constantly being changed by earthquakes and the eruption of volcanoes and by the erosion of rocks by wind and water. Deep in the interior of the geosphere, the swirling of molten rock in the outer core creates Earth's magnetic field.

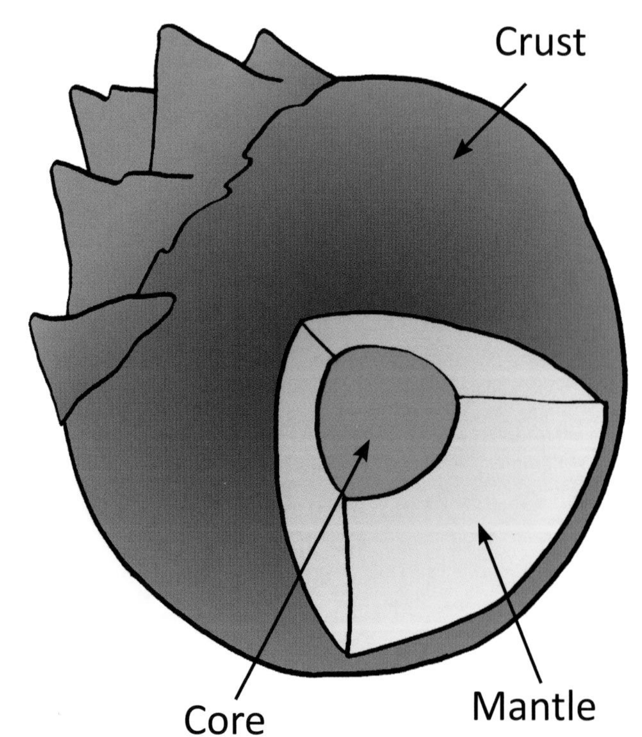

The atmosphere is made up of the air and tiny particles that surround the Earth. The atmosphere is an ever changing mixture of gases and particles that create a variety of weather events including hurricanes, clouds, cyclones, dust and sand storms, wind, typhoons, and blizzards. The atmosphere is also responsible for long-term fluctuations in climate.

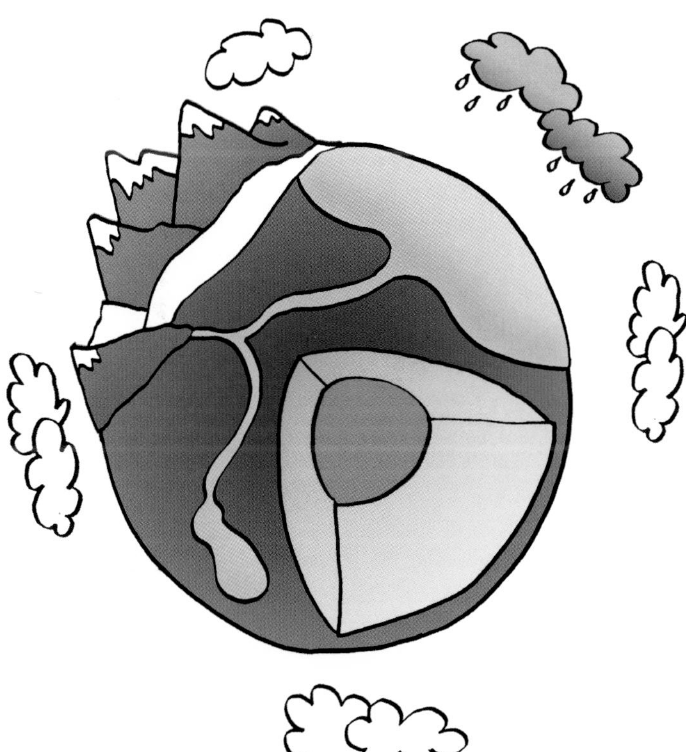

The hydrosphere makes up all the liquid and solid water on Earth. The hydrosphere includes the oceans, lakes, rivers, streams, ponds, groundwater, glaciers, water vapor in clouds, rain, snow, and icebergs. Sometimes water that is frozen and solid is considered separately from the hydrosphere and is called the cryosphere. However, in this text we will include frozen water as part of the hydrosphere.

The biosphere includes all the living things on Earth — all the life on land, in the oceans, in the dirt, underneath glaciers, and near high temperature thermal vents. All living things from the biggest land and sea creatures to the tiniest microscopic organisms are part of the biosphere.

The magnetosphere surrounds the Earth in space. The magnetosphere contains Earth's magnetic field and is formed from the interaction of the magnetic field with solar winds. The magnetosphere protects the Earth from harmful ions and solar winds coming from the Sun, and it can change shape as a result of solar variations.

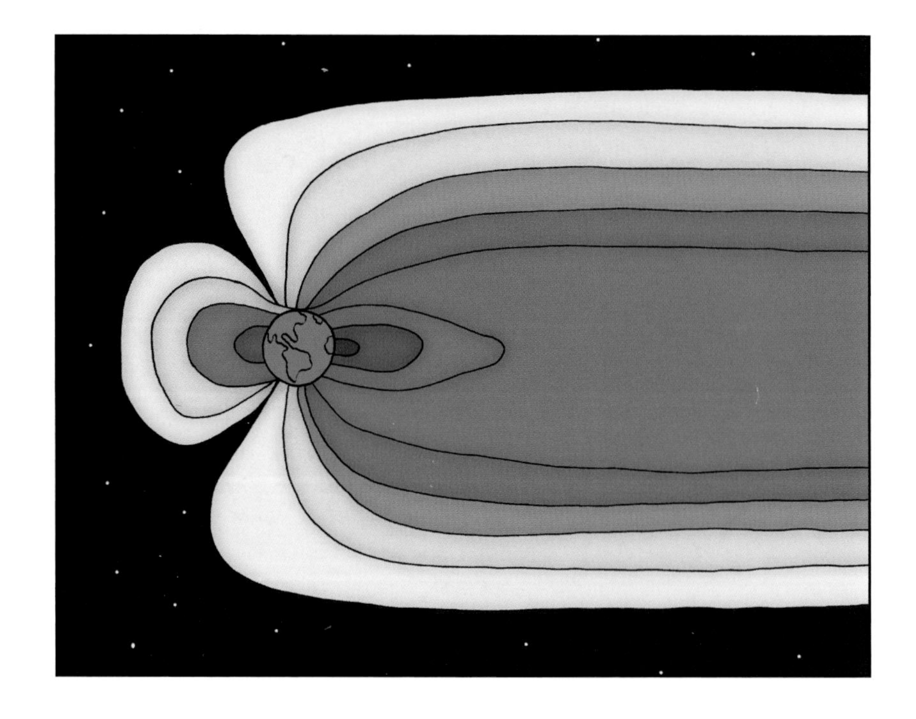

6.3 Connecting the Spheres

All of Earth's spheres are connected to each other. The atmosphere is connected to the hydrosphere, the geosphere, the biosphere, and the magnetosphere. The hydrosphere is connected to the atmosphere, the geosphere, the biosphere, and the magnetosphere. All of the spheres connect to each other. A scientist who studies climate must consider influences from the atmosphere, hydrosphere, biosphere, geosphere, and magnetosphere just as a scientist who is studying plants and animals must consider the geosphere, atmosphere, hydrosphere, and magnetosphere in addition to the biosphere.

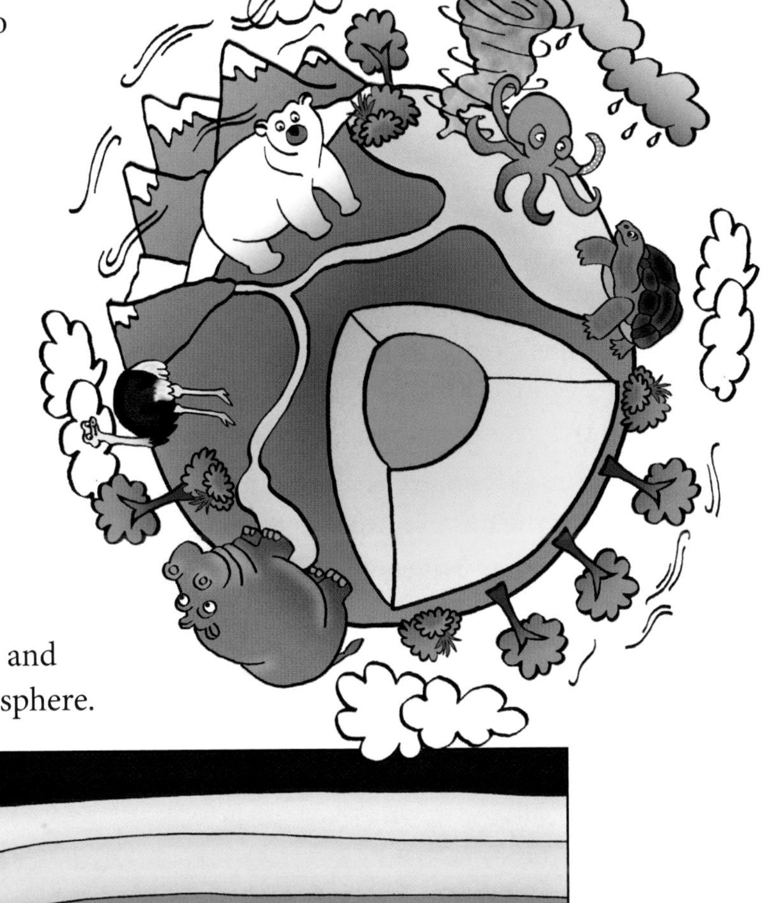

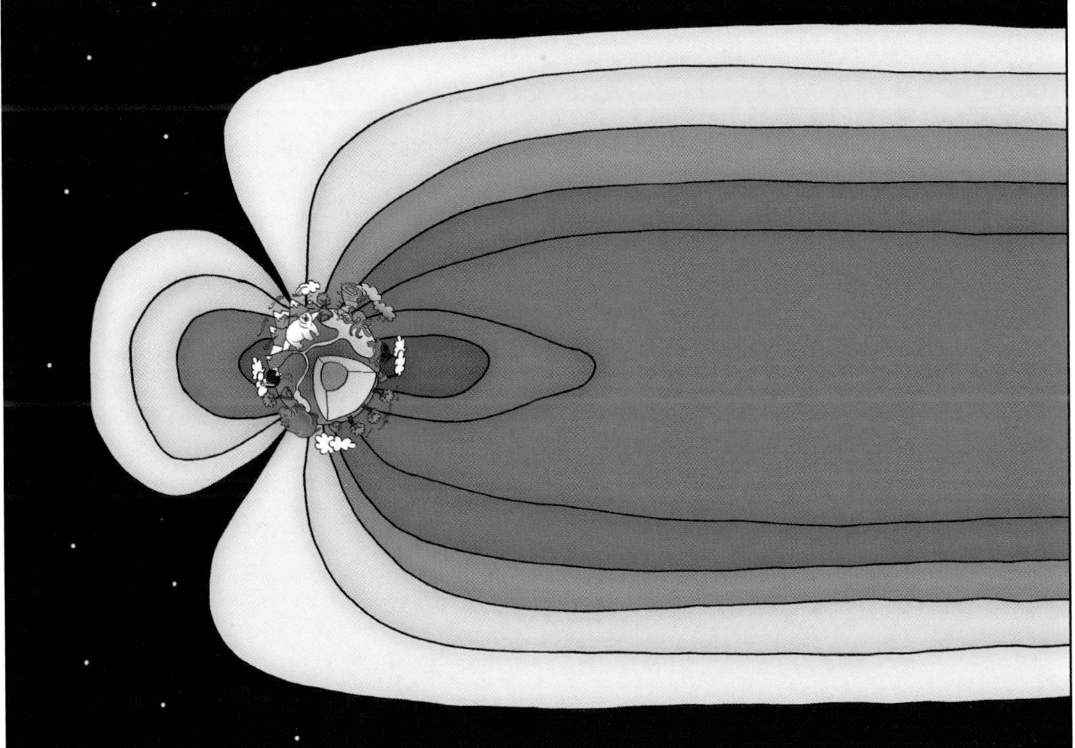

It is not always easy to determine how the spheres interact with one another. Although some effects can be immediately seen, others may take years or decades to notice. For example, a typhoon caused by the interaction of the atmosphere and hydrosphere may wipe out the entire habitat (biosphere) of one part of an island, and the change in plant and animal populations will be immediately evident. However, if that typhoon created a channel through soft soil (geosphere) that continues to erode with every rain storm, a river or gorge may develop over years or decades.

Recall from the previous section that Earth's spheres can be thought of as a system made up of different parts. The nature of Earth as a system is obvious when we notice how the different parts (spheres) interact with one another. Many of the interactions between the spheres occur in complicated cycles.

Some major cycles that connect the spheres include the energy cycle, the water cycle, the rock cycle, the carbon cycle, and the nitrogen cycle. Understanding how each of these cycles interacts with the different spheres of Earth is like putting together a huge, complicated puzzle. It takes many scientists working simultaneously on different areas of research to understand what the puzzle pieces are, how they work, and how they interact with each other.

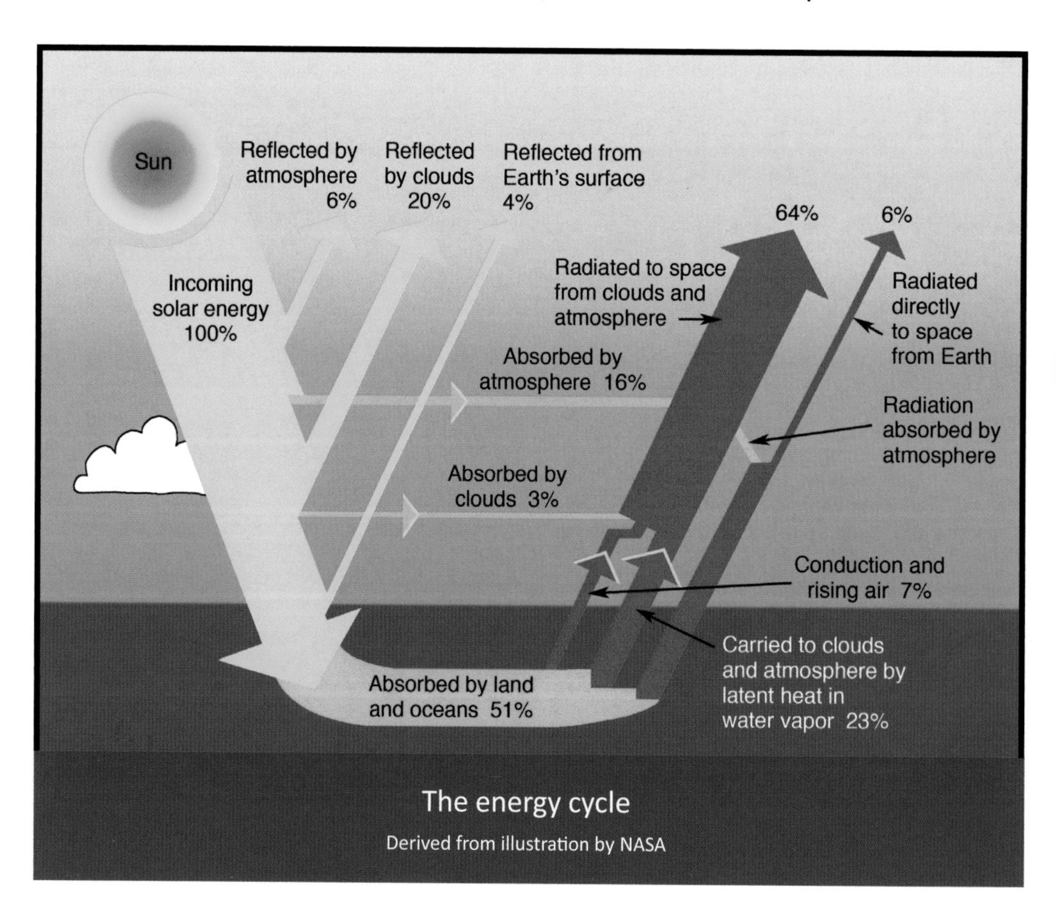

The energy cycle

Derived from illustration by NASA

To better understand how Earth's spheres interact, scientists have been taking photographs of Earth's surface from satellites. Using satellite images scientists can observe how changes in weather patterns alter coastlines, forested areas, glaciers, and waterways. Satellite imagery can also help scientists understand how changes in glaciers, forested areas, waterways, and coastlines can change the climate in different parts of the world. Also, changes to the spheres from natural disasters and human activity can be observed and analyzed.

The satellite images below show how flooding can affect the geosphere, biosphere, and hydrosphere and how fire can affect the atmosphere and biosphere.

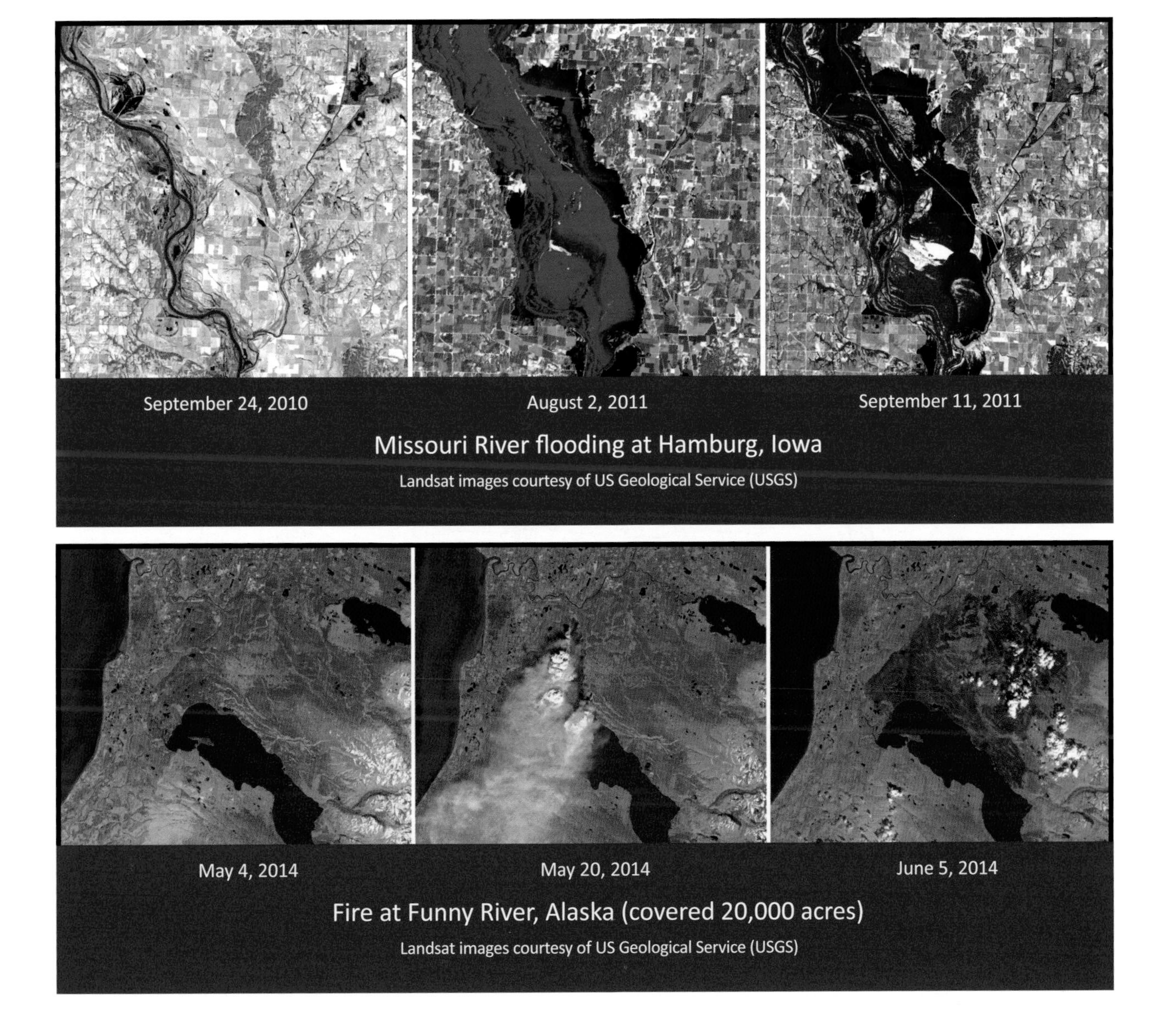

6.4 A Delicate Balance

Because Earth's spheres are connected to each other through complex cycles, there is a delicate balance between them. Earth has to maintain this balance in order for living things to grow and be healthy. Activities that disrupt one or more of these interconnected cycles can throw Earth's ecosystems out of balance.

For example, it is well know that burning fossil fuels causes air pollution in the atmosphere, lowering the pH of rain, making the rain acidic. Acid rain in the hydrosphere can cause damage to the leaves of trees and other plants in the biosphere and can make the soil of the geosphere more acidic. Acid rain can also speed up the rate of decomposition in the soil. Decomposition of leaves and other organic matter releases carbon dioxide into the soil, and the CO_2 in the soil is gradually released into the atmosphere. Rapid decomposition can result in greater quantities of carbon dioxide being produced and released from the soil into the atmosphere. This adds to the gases in the air that prevent heat from escaping from Earth, causing a warming effect which leads to further changes in the atmosphere, hydrosphere, biosphere, and geosphere.

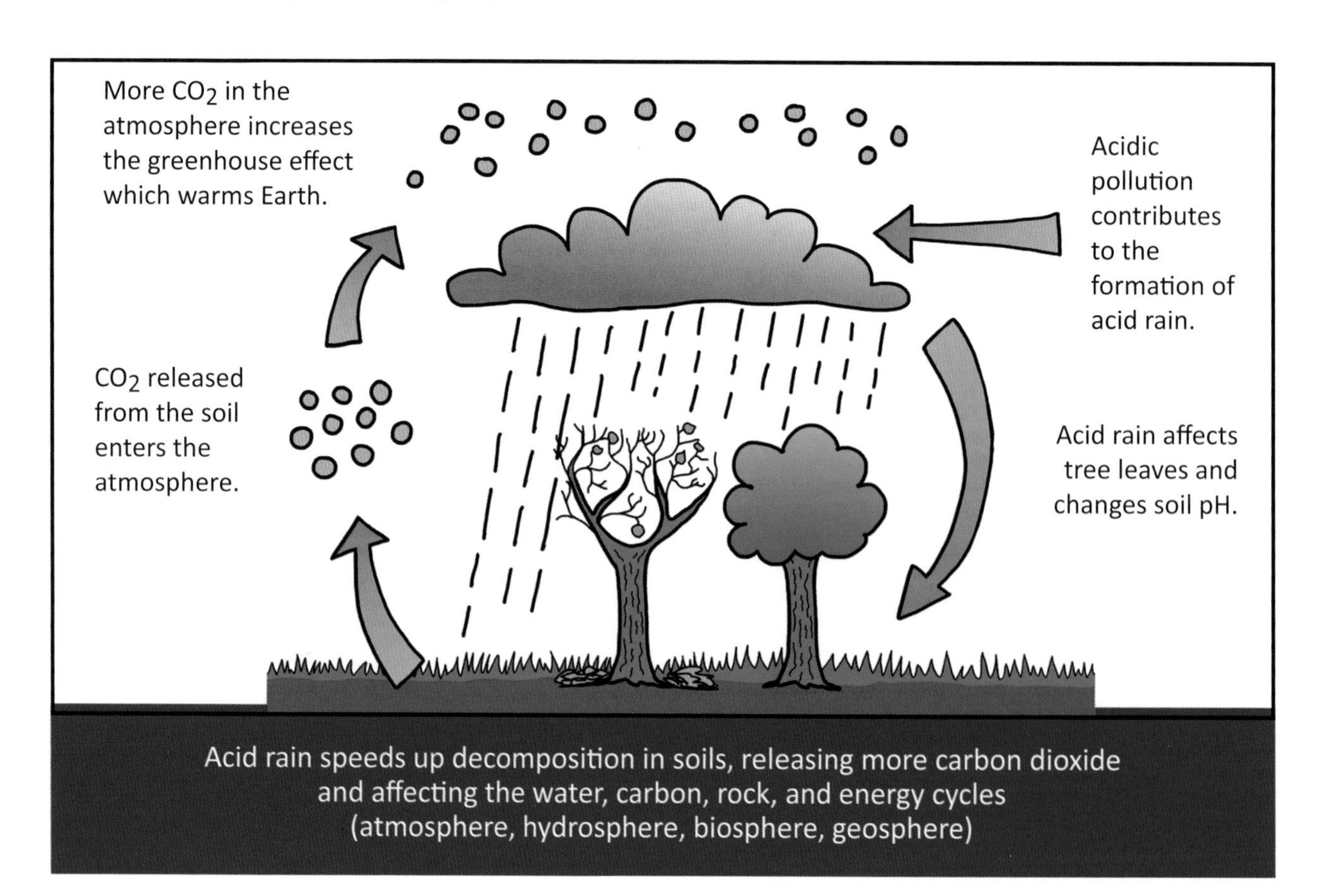

Also, natural disasters can effect the balance of Earth's spheres. A massive volcanic eruption can put so much ash in the atmosphere that the ash particles can block some of the Sun's light from the Earth, causing cooling. Increased rainfall can also result as water droplets form on ash particles. Ash and lava that fall on the geosphere can kill plants and animals and change soils and landforms.

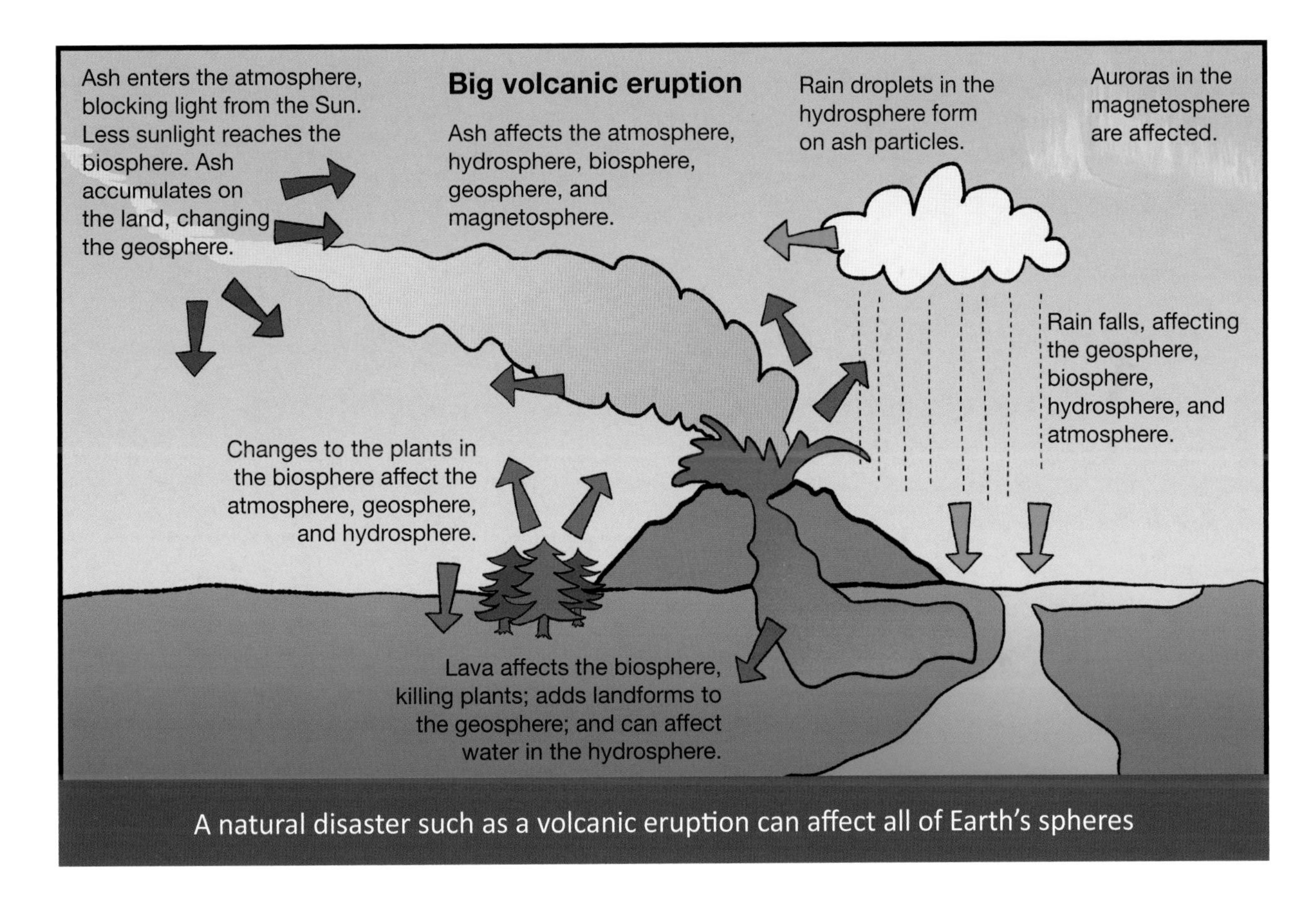

A natural disaster such as a volcanic eruption can affect all of Earth's spheres

There are many properties of Earth that must remain balanced in order for humans to thrive. Humans require a narrow temperature range, a certain amount of oxygen to breathe, a certain amount of water to drink, adequate amounts of fertile land to grow crops and raise animals for food, and materials for making clothing and building shelter.

Over the last several decades scientists have become increasingly concerned about how human activities may be disrupting the delicate balance of Earth's interconnected systems. For example, scientists have been observing changes in the amount of carbon dioxide in the atmosphere and changes in Earth's temperatures. Carbon dioxide is called a greenhouse gas when it enters the atmosphere because it affects how Earth regulates its temperature in a way that is similar to how a greenhouse regulates temperature. Earth's average temperature is the result of an intricate balance between heating by the Sun and the escape of heat from the atmosphere into space. Greenhouse gases play a role in this balance. We'll learn more about the atmosphere and greenhouse gases in Chapter 8.

Studying Earth's spheres and how they fit together and interact can help scientists come up with new ways for predicting changes to the Earth from natural and human causes and develop new ideas for helping people protect and restore environments.

6.5 Summary

- Earth can be described as a system of interconnected spheres that include weather, plants and animals, rocks and dirt, water, energy, and Earth's magnetic field.
- Earth's spheres are commonly referred to as: the atmosphere, the hydrosphere, the geosphere, the biosphere, and the magnetosphere.
- Earth's spheres are interconnected, and changes in one sphere can result in changes in another sphere.
- Earth must maintain a delicate balance between the spheres to continue to support life.

6.7 Some Things to Think About

- Take your field notebook and go for a hike. Observe how the different parts of Earth work together: air and clouds, water, soil and rocks, plants and animals, and the Sun.
- Which sphere do you think is the most important to life on Earth? Why?
- Describe some examples of how you have seen Earth's spheres interact. Which spheres were involved and what took place?
- Do you think there are natural events that could make Earth's spheres become unbalanced? Why or why not? If so, do you think after a time balance would be restored? Why or why not?
- What ideas do you have for helping to protect and restore environments? Why do you think this is important?

Chapter 7 The Geosphere

7.1 Introduction

If you take a walk through the forest, you might notice birds sitting in the trees, small plants growing from the soil, mushrooms bursting out of rotten logs, and squirrels gathering food for the winter. All of the living things you see are supported by the rocks, soil, and minerals beneath the forest floor. Without a solid surface, plants could not grow and animals could not collect the food they need for survival. The rocky part of Earth is called the geosphere and is made of rocks, minerals, soils found on the surface, and molten rock, or magma, found deep below the surface.

Recall that humans live on the crust, the uppermost part of the geosphere. Although people can travel in the sky to get to another city or traverse the oceans in a boat, all of human life is supported by the rocky geosphere. Humans build houses from materials found on Earth's crust, grow plants and raise animals on Earth's crust, and use rocks and minerals from the crust to create cars, airplanes, and trains. Humans also extract gas and coal from the crust to fuel factories and power boats and even use nuclear energy obtained from rocks to provide electricity to entire cities. We know a lot about the crust, but what do we know about the parts of Earth below the crust? How do geologists study the other layers of Earth?

7.2 Using Volcanoes To See Inside Earth

How do scientists know what Earth's interior is made of if we can't sample rocks and minerals below the upper part of the crust? One way scientists study rocks and minerals deep in the Earth is to observe what happens when volcanoes erupt and to study the lava, rocks, and ash that come from the volcanoes.

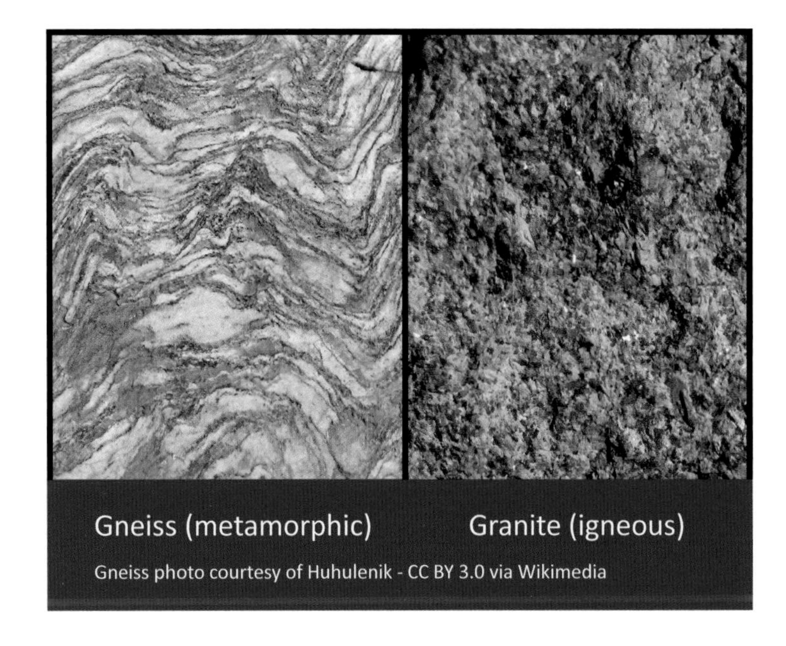

Gneiss (metamorphic) Granite (igneous)

Gneiss photo courtesy of Huhulenik - CC BY 3.0 via Wikimedia

In a volcanic eruption, magma from within the Earth is brought to the surface. In addition, pieces of rocks formed inside the Earth are often ejected when they are torn loose and carried by the magma as it pushes towards the surface with tremendous force. Granite, an igneous rock, and gneiss, a metamorphic rock, are the most common types of rock ejected by volcanoes.

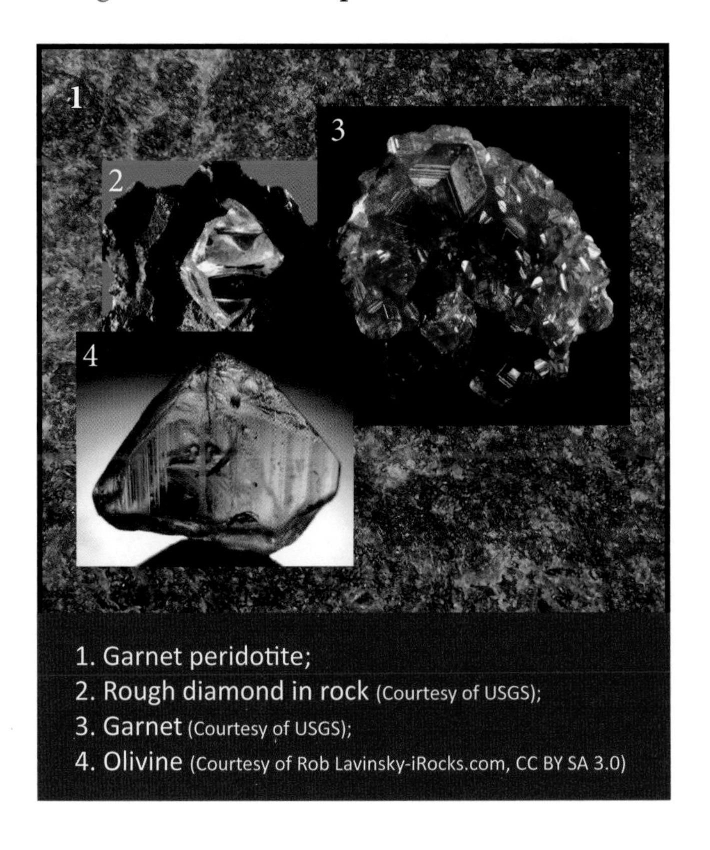

1. Garnet peridotite;
2. Rough diamond in rock (Courtesy of USGS);
3. Garnet (Courtesy of USGS);
4. Olivine (Courtesy of Rob Lavinsky-iRocks.com, CC BY SA 3.0)

Rare gems such as diamonds have been found in rocks that have been ejected during volcanic eruptions. Diamonds are made of pure carbon and most often form more than 150 km (93 mi.) below Earth's surface. Volcanoes also bring up chunks of peridotite, an igneous rock made mostly of the minerals olivine and pyroxene. Peridotite may also contain garnets, some of which can be used as gemstones. Laboratory experiments show that garnet peridotite forms at depths greater than 50 km (31 mi.).

Volcanoes and Volcanologists

1. A volcanic eruption in Hawaii Courtesy of Hawaiian Volcano Observatory (HVO)/USGS by J. D. Griggs
2. Setting up a GPS system to measure deformation of ground (becoming higher or lower), Mt. St. Helens, Washington State Courtesy of USGS by Mike Poland; 3. Collecting volcanic ash samples in Alaska Courtesy of Alaska Volcano Observatory (AVO)/USGS by Kristi Wallace; 4. Taking a sample of lava, Kilauea, HI Courtesy of HVO/USGS; 5. Taking gas samples at the Cookie Monster skylight, HI Courtesy of HVO/USGS by J. D. Griggs; 4. Collecting lava samples, Kilauea, HI (hornito formation in background) Courtesy of HVO/USGS

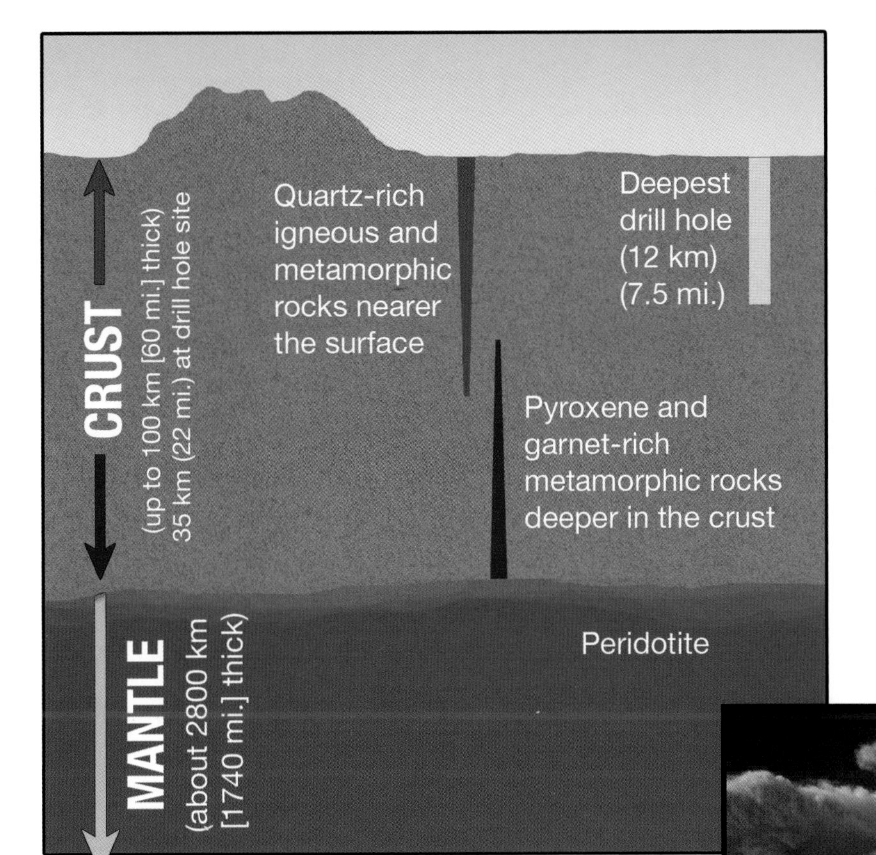

Information gathered from volcanoes has given scientists an idea about what forms below Earth's surface, with quartz-rich rocks forming nearer the surface of the crust and pyroxene and garnet peridotite in the lower part of the crust closer to the mantle.

Geologists also study the makeup of volcanic ash to learn more about the interior structure of Earth. Volcanic ash is formed when a volcano erupts explosively. Magma contains dissolved gases that can cause the magma to explode if the gases expand and escape violently into the atmosphere. Solid rock can also be exploded when the force of the escaping gases shatters the rocks.

Ash from Mt. St. Helens, WA collected 39 km away in Idaho
Courtesy of Cascades Volcano Observatory/USGS

Eruption plume of gases and ash Mt. St. Helens, Washington State
Courtesy of Cascades Volcano Observatory/USGS

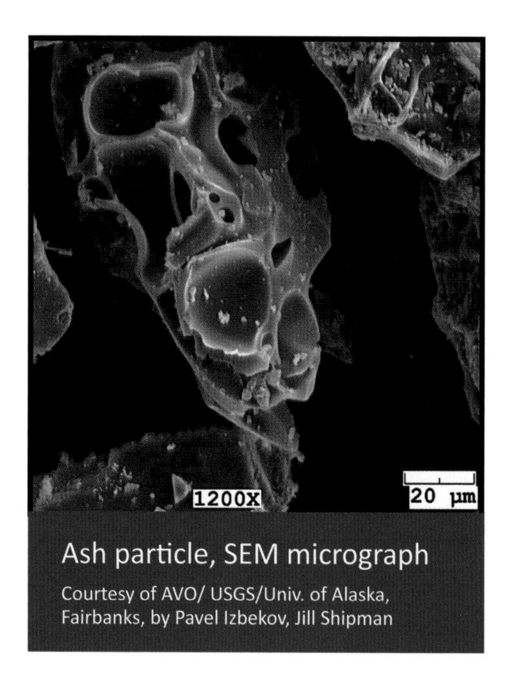

Ash particle, SEM micrograph
Courtesy of AVO/ USGS/Univ. of Alaska, Fairbanks, by Pavel Izbekov, Jill Shipman

Volcanic ash isn't soft like the ash from a wood fire. Instead, it is made of hard, sharp bits of rock that are the size of grains of sand or smaller. Because the force of the explosion thrusts the ash up into the atmosphere, winds can carry the tiny particles far from the volcano — as much as thousands of kilometers away! By studying the composition of the ash particles, geologists can learn more about what the interior of Earth is made of.

7.3 Using Earthquakes To See Inside Earth

If you live in an area of the globe that's prone to earthquakes, you might have awakened in the middle of the night to a deep rumbling sound and felt your house being shaken by an earthquake. Earthquakes happen in different parts of the world and can be devastating for people living near the origin, or epicenter, of the earthquake. They can cause extensive damage to homes, schools, city buildings, farmland, and other structures that exist on Earth's surface.

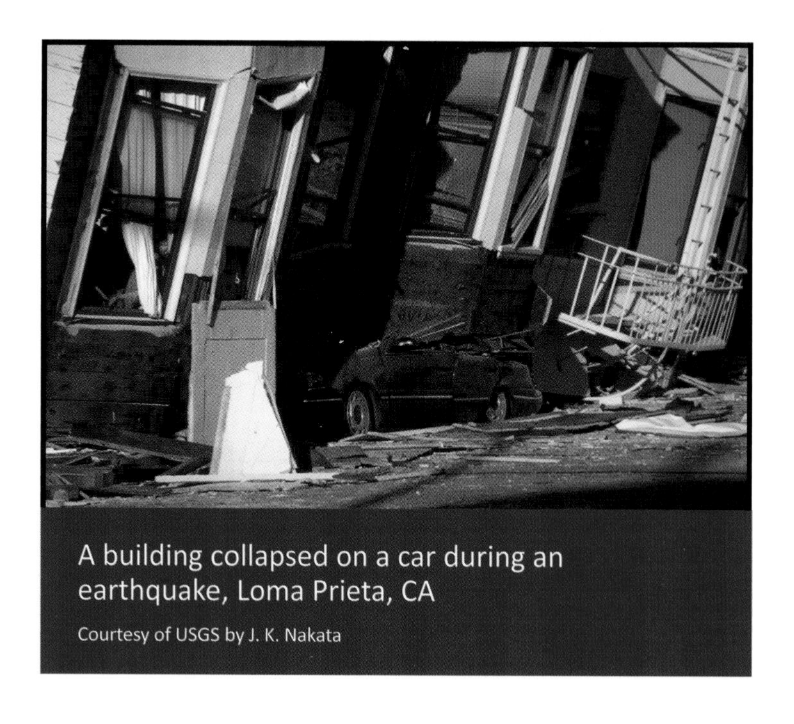
A building collapsed on a car during an earthquake, Loma Prieta, CA
Courtesy of USGS by J. K. Nakata

However, earthquakes are very useful for finding out more about Earth's layers. Earthquakes create waves that are caused by the movement of the crust and lithosphere, and these waves can be used to create images of Earth's interior.

Aerial view of the San Andreas Fault in California

Courtesy of Ikluft, CC BY SA 3.0

A satellite view of the Piqiang Fault in China showing how Earth's crust has moved

Courtesy of NASA Earth Observatory images by Robert Simmon and Jesse Allen, using Landsat data from the USGS Earth Explorer

Waves are the transfer of energy through a material such as air, water, wood, rocks, or magma. When you throw a pebble into a pond or strike a tuning fork against a surface, you are creating waves.

When the tectonic plates of Earth's crust move, slide, or bump up against each other, stresses build up as the rough edges of the plates stick together, but the plates themselves continue to move. An earthquake occurs when the stresses become too great and the plates suddenly lurch as the edges release their hold on each other. An earthquake starts at the point where the release between two plates occurs (the epicenter), and the energy this sudden movement creates travels through the surrounding rock in the form of waves.

To measure the waves generated by an earthquake, geologists use a seismometer. Recall from Chapter 2 that a seismometer is a very sensitive instrument that can detect and amplify waves on Earth's surface, in the body of the crust, and below the crust.

How can earthquake waves tell us about Earth's interior? It turns out that the way an earthquake wave travels through the Earth's crust and also through Earth's lower layers allows geologists to construct an image of Earth's interior. Earthquake waves can be used like an x-ray, giving scientists a picture of Earth's internal structure. By using seismometers stationed in many locations on Earth, scientists can determine how fast a seismic wave from an earthquake travels through the Earth from the epicenter of the earthquake to where it comes back to the surface in another location. Scientists can also determine the direction the wave is traveling. Combining the direction of the wave with its speed reveals the velocity of the wave.

This illustration shows how two different wave patterns travel through Earth from the epicenter of an earthquake. The initial waves are caused by movements of tectonic plates that create an earthquake. The waves from the earthquake then pulse through Earth's layers. As the waves move, they are reflected and refracted when they encounter different materials. In some cases the waves are not able to travel through a material. This leaves areas where waves cannot be detected at the surface by seismometers, and these areas are called shadow zones.

We can see that the behavior of seismic waves helps scientists explore and identify the interior parts of Earth's crust, mantle, and core. There is much yet to be discovered about Earth's interior.

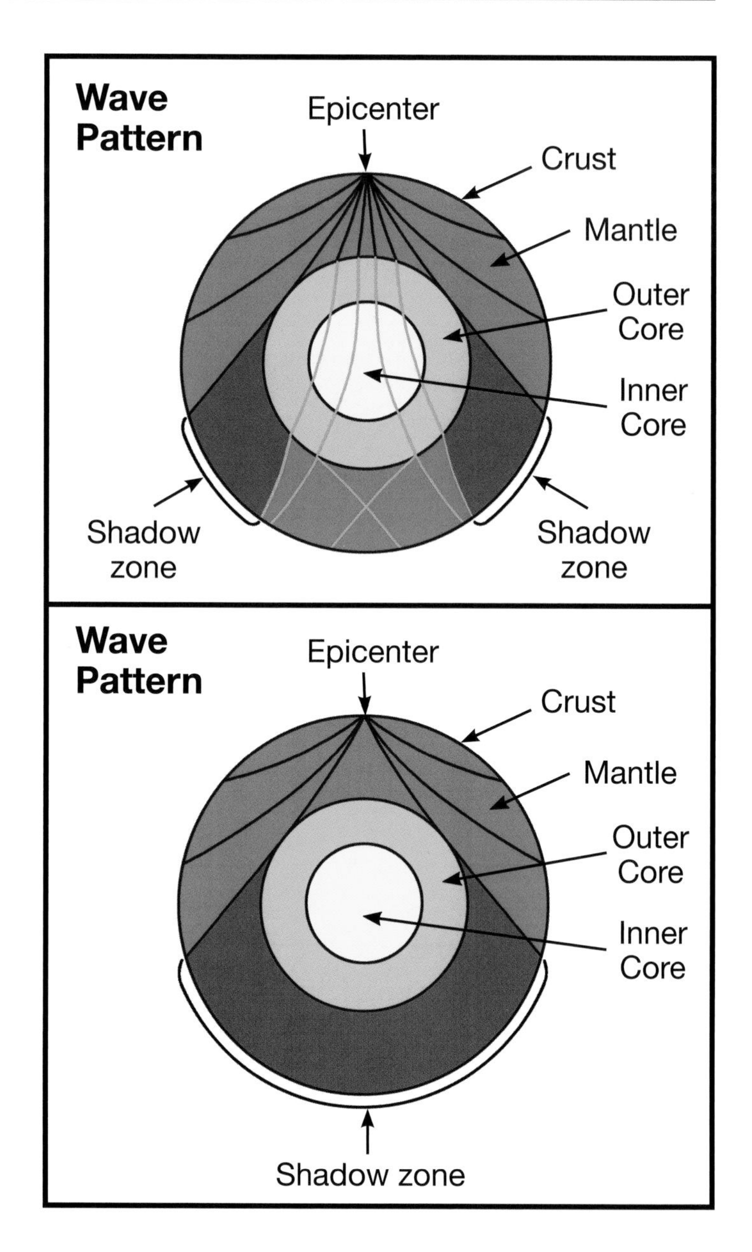

7.4 How Hot Is the Core?

How hot is the center of the Earth? If you've ever seen a volcano erupt and watched as hot lava consumed trees, buildings, and cars, you can imagine that the molten rock in the mantle of the Earth is very hot. But how hot is the molten rock at the center of Earth?

The information gathered from earthquake seismic waves traveling through Earth gives scientists some very basic information about the temperatures of the inner and outer cores. Because the analysis of seismic waves can determine whether rock is solid or liquid, scientists can tell whether the rock is above or below its melting temperature. As we have seen, the movement of seismic waves suggests that the outer core is liquid and the inner core is solid.

Since the inner core is thought to be mostly iron that is under extreme pressure from the weight of all the materials above it, lab tests have been done that put iron under extreme pressure to get an indication of what core temperatures might be. The results of these tests combined with computer modeling have helped scientists arrive at estimated temperatures for the inner core.

Currently, the combined data gathered by various research methods suggests that the outer core is about 3000° to 5000°C (5400° to 9000°F). The inner core is thought to reach temperatures of about 6000°C (10,800°F), which is as hot as the surface of the Sun!

7.5 Summary

- The geosphere is the rocky part of Earth and is made of rocks, minerals, soils found on the surface, and magma found deep below the surface. The geosphere extends all the way from the surface of Earth to its very center.
- The Earth has layers that are divided into different sublayers.
- Lava, ash, and rocks ejected by volcanoes can be used to investigate what exists below Earth's crust.
- Earthquakes create waves that can be detected by seismometers and used to investigate Earth's layers.
- The temperature of Earth's inner core is believed to be about 6000°C.

7.6 Some Things to Think About

- Make a list of some of the functions of the geosphere.
- What do you think geologists learn from volcanoes?
- Do you think volcanic eruptions change the geosphere? Why or why not?

- When you have been out walking or riding in the car, what have you observed about rocks and landscapes in your region?

  Do all rocks look alike? Are they the same size, shape, color, and texture?

  Do rocks in a small area tend to look similar? What about in a large area?

  Do they form mountains or sand?

  Write about all the things you have noticed about rocks.

- What do you think earthquakes tell geologists about the geosphere?
- How do geologists study earthquakes?
- Do you think we will someday be able to predict earthquakes? Why or why not? How would earthquake prediction be helpful?
- What factors do you think would need to be taken into account when calculating the temperature of the core?

Chapter 8 The Atmosphere

8.1 Introduction

We know that we need to have air to breathe in order to live. But what is air made of? How far does it extend above the surface of the Earth? Is all of the air the same? Why doesn't it float away into space?

The thin layer of gases that surrounds the Earth is called the atmosphere. The word atmosphere comes from the Greek word *atmos*, which means "vapor" and refers to the gaseous state of matter. Without the atmosphere, life on Earth would not be possible. Earth's atmosphere contains the oxygen that animals need to breathe in order to live. It also protects the Earth and its inhabitants from getting too much energy from the Sun and helps keep the Earth's temperature from getting too hot or too cold. It carries the water that falls as rain and snow.

8.2 Chemical Composition

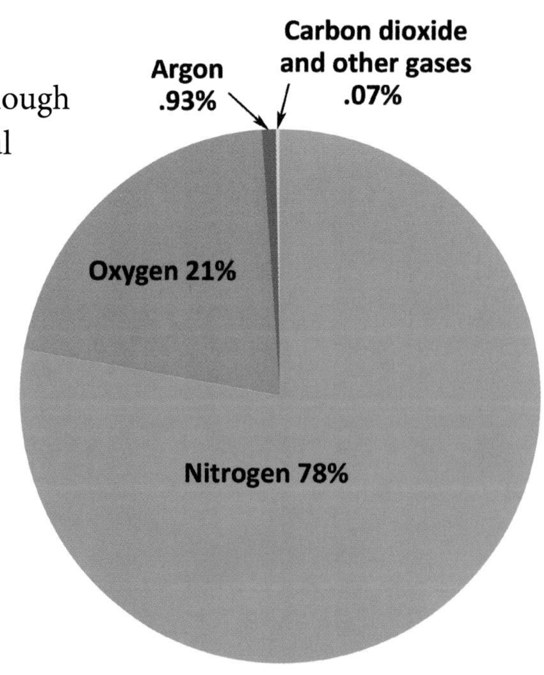

Earth's atmosphere is made up of several gases. Although oxygen is essential for life, it is not the most plentiful gas in the atmosphere. Instead, nitrogen makes up about 78% of the atmosphere, with oxygen at about 21%, and other gases including carbon dioxide, ozone, methane, and argon making up the remaining 1%. These percentages are for the "dry" atmosphere. But water vapor is also present in the atmosphere at an average of about 1% of volume.

There are several features in Earth's atmosphere that allow life to exist and also protect it. These features include the amount of carbon dioxide, the ozone layer, water vapor, and the presence of oxygen.

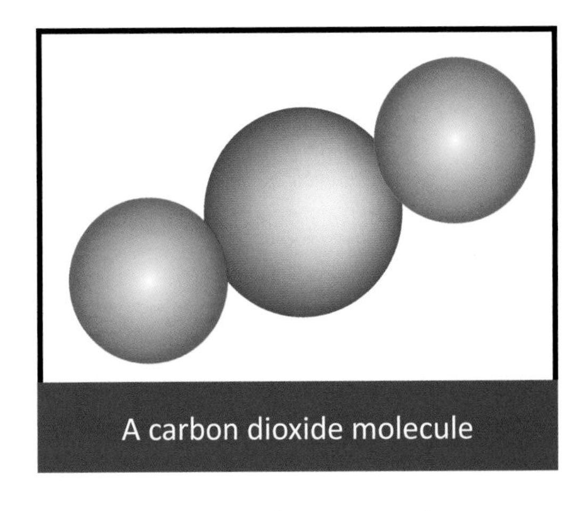
A carbon dioxide molecule

Carbon dioxide (CO_2) is a molecule made of one carbon atom and two oxygen atoms. It makes up less than 1% of the atmosphere. Carbon dioxide helps stabilize Earth's climate by trapping some infrared energy (heat) from the Sun which keeps Earth from freezing. However, too much carbon dioxide can trap too much infrared energy which can then cause the Earth to get warmer (see Section 8.6). Carbon dioxide is also essential to the process of photosynthesis in plants. Plants convert carbon dioxide to sugars to use for food. They then release oxygen into the atmosphere. Photosynthesis is the major source of oxygen on Earth.

Natural sources of atmospheric carbon dioxide include animal and plant respiration, by which oxygen and nutrients are converted into carbon dioxide and energy; ocean-atmosphere exchange, in which the oceans absorb and release carbon dioxide at the sea surface; and volcanic eruptions, which release carbon from rocks deep in the Earth's crust.

Man-made sources of carbon dioxide include the burning of fossil fuels for heating buildings, power generation, and transportation, as well as some industrial processes.

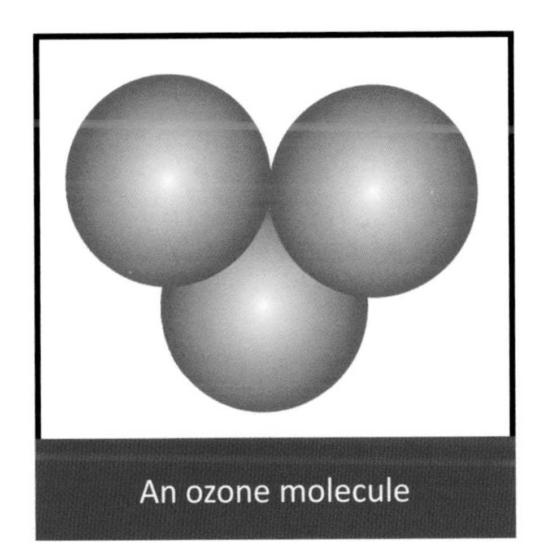
An ozone molecule

Ozone (O_3) is a molecule made of three oxygen atoms. Ozone is also an important gas in our atmosphere although it makes up less than .0001% of the total atmospheric gases. The ozone layer blocks most of the ultraviolet radiation from reaching the Earth's surface (see Section 8.3) but allows some of the incoming energy from the Sun to be trapped in the atmosphere. This process also slightly heats the Earth, keeping temperatures mild and stable. Without the ozone layer, Earth would be uninhabitable.

Water vapor is also an important atmospheric gas. The atmosphere close to the Earth's surface contains up to 4% water vapor. This water vapor greatly affects our weather. Water vapor is actually the most important molecule for keeping our Earth warm. Water vapor is also responsible for the formation of clouds and rain that give plants and animals the essential water they need for life.

Finally, the most important atmospheric gas for animal life is oxygen. Without oxygen, life as we know it would not be possible. Animals breathe in oxygen, and this oxygen is moved through cells and used for a variety of biochemical processes that make animal life possible.

8.3 Structure of the Atmosphere

Most of the mass of Earth's atmosphere exists within 20 kilometers (12 miles) of the Earth's surface. However, above these first 20 kilometers of air there are several atmospheric layers that function to protect the Earth's surface, absorb harmful radiation, and keep the Earth warm.

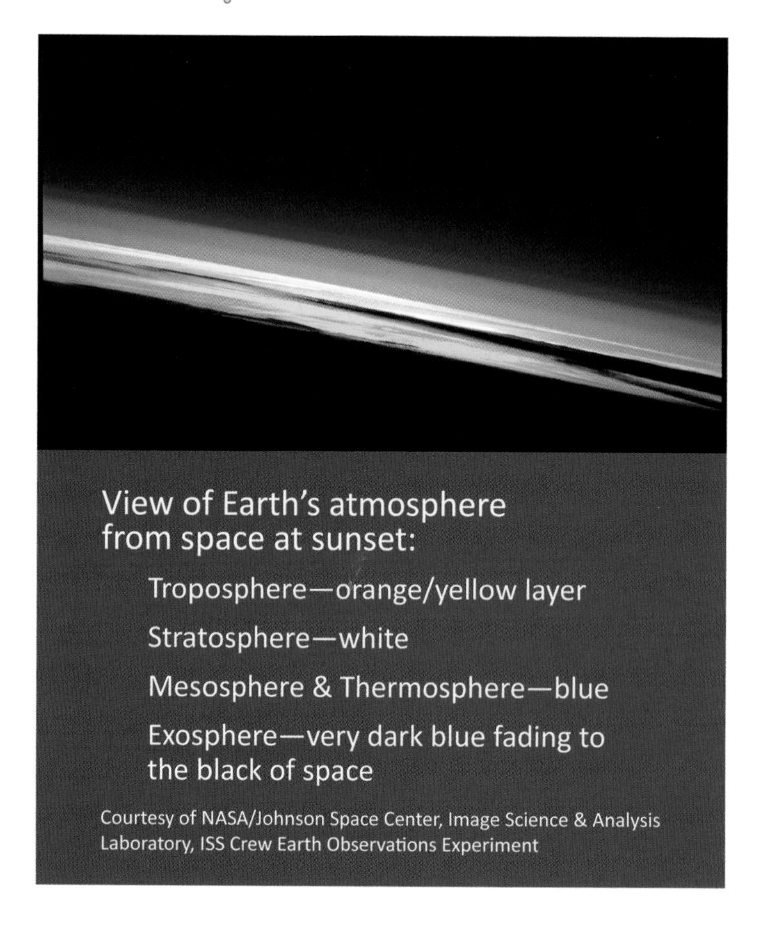

View of Earth's atmosphere from space at sunset:

Troposphere—orange/yellow layer

Stratosphere—white

Mesosphere & Thermosphere—blue

Exosphere—very dark blue fading to the black of space

Courtesy of NASA/Johnson Space Center, Image Science & Analysis Laboratory, ISS Crew Earth Observations Experiment

Scientists have divided the atmosphere into five layers — troposphere, stratosphere, mesosphere, thermosphere, and exosphere. These different layers are the result of differences in temperature, chemical composition, movement, and density of the gases in the atmosphere.

The top boundary of each atmospheric layer is called a "pause," so there is a tropopause, a stratopause, a mesopause, and a thermopause. A "pause" is an area where a change in temperature occurs, with the temperature either changing abruptly or reaching a minimum or maximum temperature.

The first layer above the Earth's surface is called the troposphere and is the layer we live in. The troposphere contains 78-80% of the Earth's atmospheric gases and extends to a height of about 6-20 kilometers (4-12 miles) from the Earth's surface. All of the weather we experience happens in the troposphere.

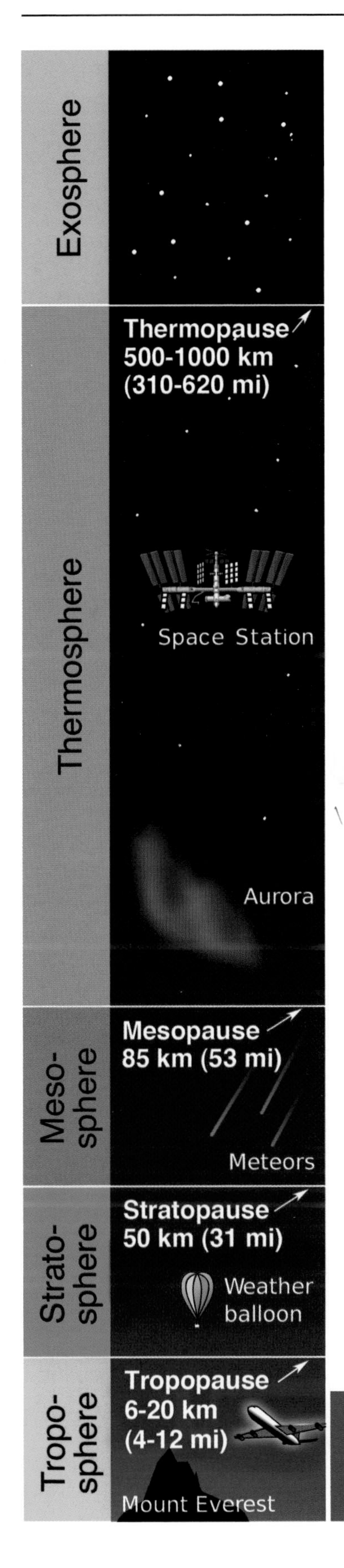

The layer above the troposphere is called the stratosphere. The stratosphere extends approximately 50 kilometers (31 miles) above Earth's surface. The air in the stratosphere is very dry with few gases and very little water vapor. Because there are fewer gases, this layer is more stable than the troposphere. Few, if any, clouds form in the stratosphere and airplanes fly in the lower stratosphere to avoid weather conditions that cause turbulence and bumpy flights.

The ozone layer is found at the lower part of the stratosphere. Ozone is crucial to life on Earth because ozone absorbs harmful ultraviolet radiation from the Sun. Ultraviolet radiation has a shorter wavelength than visible light and can damage living cells. Sunscreen and sunglasses help protect people from the damage ultraviolet radiation causes. Without the ozone layer, organisms would suffer severe burns from ultraviolet radiation. The ozone layer also creates increased temperatures in the stratosphere.

The mesosphere begins at an altitude of approximately 50 kilometers (31 miles) and reaches heights of about 85 kilometers (53 miles). In this layer the temperatures again begin to get cooler the higher you go. The top of the mesosphere is the coldest part of the atmosphere with temperatures of approximately -90° Celsius (-130° Fahrenheit).

Most meteors that enter the Earth's atmosphere don't make it through the mesosphere. Because they are traveling at extreme speeds, they will break apart and burn up before reaching the Earth's surface.

Structure of the Atmosphere

Derived from illustration by National Weather Service, National Oceanic and Atmospheric Administration (NOAA)/Department of Commerce

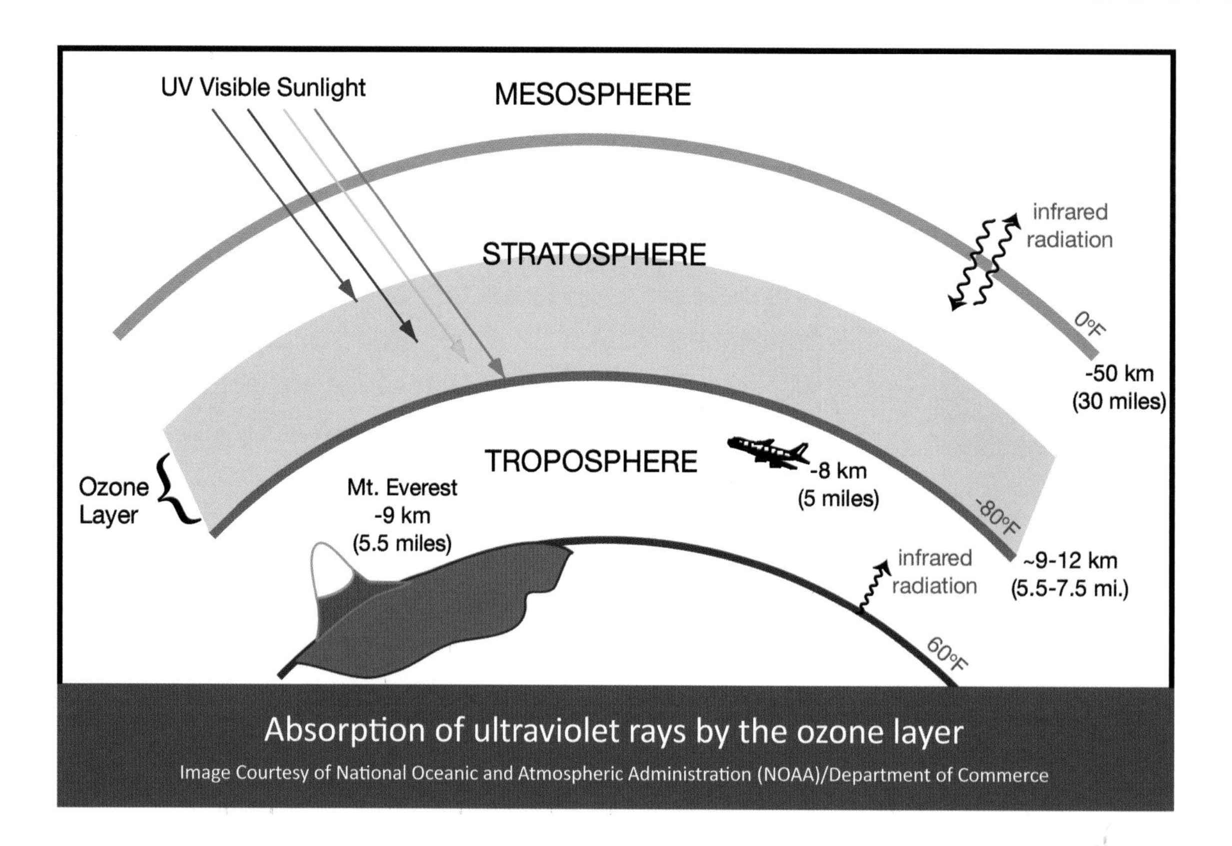

Absorption of ultraviolet rays by the ozone layer

Image Courtesy of National Oceanic and Atmospheric Administration (NOAA)/Department of Commerce

The next layer above the mesosphere is the thermosphere. The top of the thermosphere can reach altitudes ranging from about 500-1,000 kilometers (310-620 miles). This varying range in height is caused by temperature changes resulting from heat coming from the Sun. The amount of energy released by the Sun varies, and when there is more energy released, temperatures rise, causing the thermosphere to expand. With less energy released, temperatures drop, and the thermosphere contracts, getting smaller. Temperatures also vary significantly between day and night when the Sun is either shining or not shining on a part of the Earth.

The International Space Station orbits within the thermosphere, and auroras (northern and southern lights) occur in this layer.

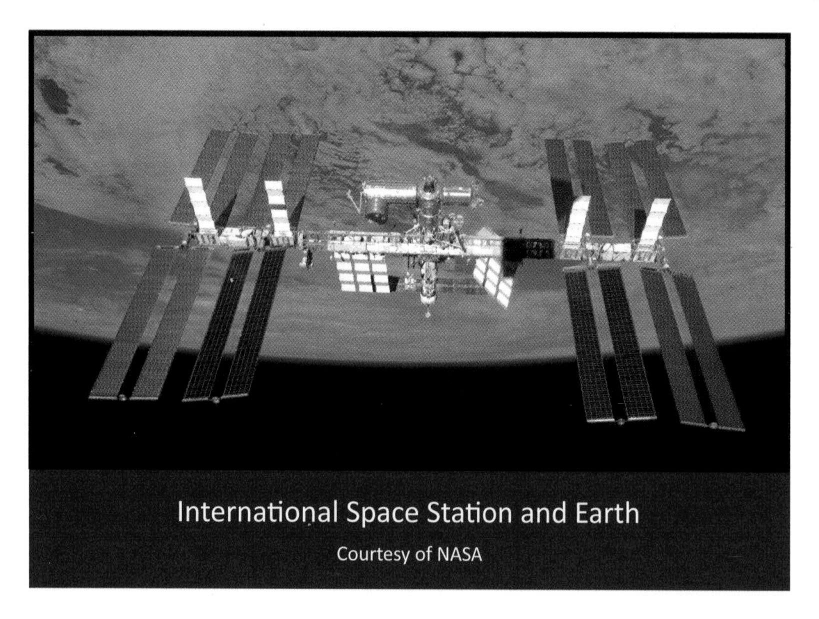

International Space Station and Earth

Courtesy of NASA

The exosphere is the outermost layer of the atmosphere and is sometimes considered to be part of space. In this layer, atoms and molecules escape into space freely.

8.4 Atmospheric Pressure

Atmospheric pressure (air pressure) is the force exerted on us by the weight of air molecules in the Earth's atmosphere. It is defined as the force per unit area exerted against a surface by the weight of the air above that surface.

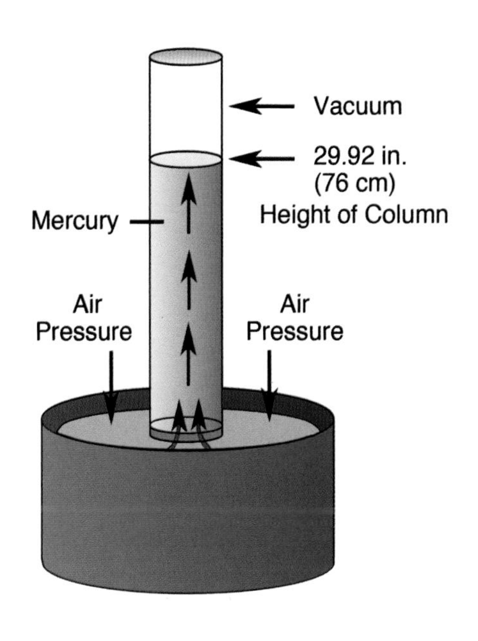

Scientists measure atmospheric pressure with barometers, which are instruments that record the air pressure in millibars (mb) or in inches of mercury. The standard air pressure at sea level is 1,013.25 millibars (29.92 inches of mercury), which is the equivalent of 6.7 kilograms (14.7 pounds).

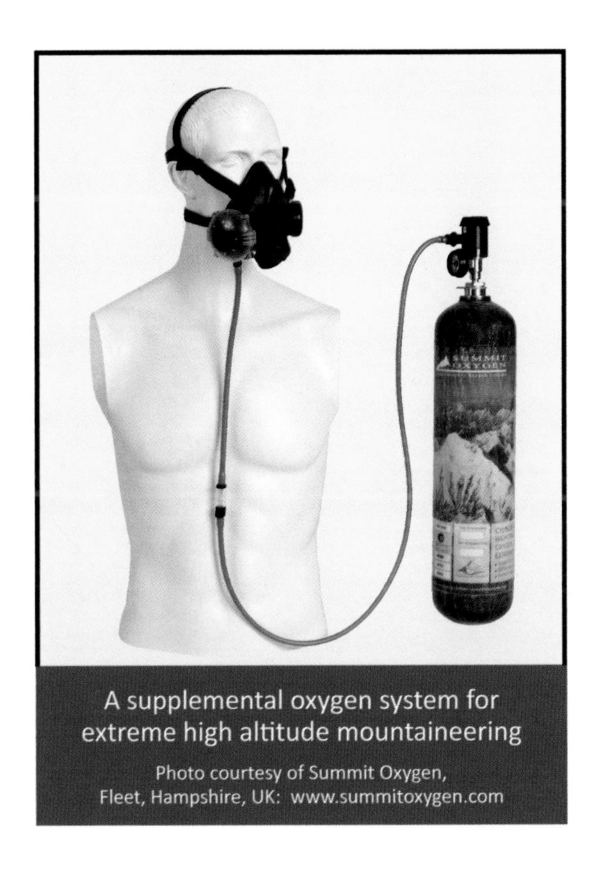
A supplemental oxygen system for extreme high altitude mountaineering

Photo courtesy of Summit Oxygen, Fleet, Hampshire, UK: www.summitoxygen.com

At sea level the air is warm, there is plenty of oxygen to breathe, and the atmospheric pressure is high. However, if you were to climb a mountain and go up in elevation, away from sea level, you would find that the temperature starts to cool and there is less oxygen to breathe and less atmospheric pressure. If you were to climb a really high mountain at very high altitudes, you would find that the temperature is even colder, there is even less oxygen to breathe, and much less atmospheric pressure.

Why does this happen? As you go from sea level to higher and higher elevations, the density of atmospheric gases (the number of gas molecules per volume) decreases, resulting in less air pressure. With less air pressure and fewer gas molecules, the remaining molecules will expand into the available space. As it turns out, when gases expand, they cool. This is why mountain climbers must carry lots of warm clothing and sometimes tanks of oxygen when they climb high mountains.

Mount Everest is the highest place on Earth with an elevation of 8,848 meters (29,029 feet). At the top of the mountain the average air pressure is about 300 millibars. This means that there is only about one-third of the oxygen that there is at sea level.

8.5 Gravity and the Atmosphere

Several factors keep Earth's atmosphere from floating away into space. These factors include the size, or mass, of the gas atoms and molecules, gravity, and the escape velocity. The escape velocity is the minimum speed needed for an object to escape from the gravitational field of a star (sun), planet, or moon. Earth's escape velocity prevents atoms and molecules in the atmosphere from overcoming the force of gravity and escaping into space.

Dawn spacecraft is launched to attain escape velocity
Courtesy of NASA/Sandra Joseph, Rafael Hernandez

Some other celestial bodies in our solar system do not have atmospheres. The Moon, for example, has such a small atmosphere that it is considered negligible. This is due to the fact that the Moon is less massive than Earth and therefore has a weaker gravitational force. This weaker gravitational force creates a smaller escape velocity for the Moon. Earth's escape velocity is 11 kilometers per second. The Moon's escape velocity is 2.4 kilometers per second. Therefore, it is much easier to escape into space from the Moon than from Earth. Although there is gravity on the Moon, it is too weak to hold a significant atmosphere.

8.6 The Greenhouse Effect

Greenhouses are enclosed structures made of glass or plastic. The glass or plastic lets sunlight enter and then traps some of the heat from the Sun, holding it inside the greenhouse. Trapping the heat keeps a greenhouse warm inside even during winter, allowing plants to grow year round. Our planet's atmosphere traps energy from the Sun in a similar way, and this is called the greenhouse effect. The greenhouse effect occurs because energy from the Sun enters Earth's atmosphere, but not all of it escapes back into space. Some gases act like a greenhouse by absorbing heat and keeping Earth warm. These gases include carbon dioxide, water vapor, and methane, among others.

During the day, sunlight warms the planet. At night the surface of the Earth cools and releases heat back into the atmosphere. Some of this heat does not escape into space but is trapped by greenhouse gases. If this process remains balanced, the Earth will stay at about

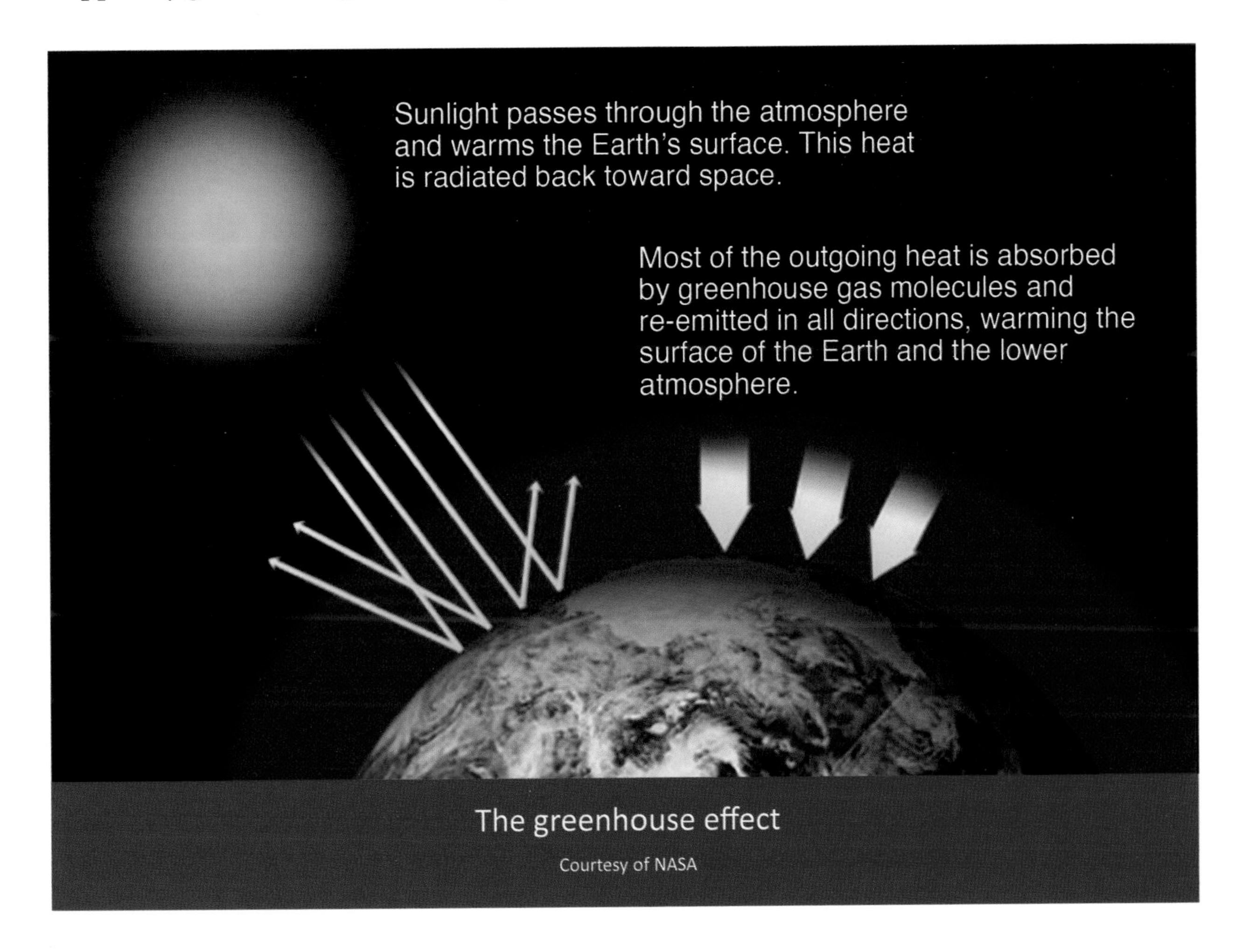

The greenhouse effect

Courtesy of NASA

the same yearly average temperature. If the greenhouse effect is too strong, it won't allow as much heat to escape into space, and then Earth begins to get warmer.

Smoke from burning fossil fuel

Courtesy of National Oceanic and Atmospheric Administration (NOAA)/ Department of Commerce

Every day scientists collect a lot of data about air and ocean temperatures all over the world. The data show that currently there is a rise in the average temperature of the Earth's atmosphere and oceans. Some indications of a warming of Earth are record high temperatures at different locations, the melting of glaciers, severe flooding, droughts, and fires. Although the warming trend is clear, the causes of the warming are under debate.

Measurements show that there is an increase of greenhouse gases in the atmosphere. Most scientists believe that human activity is a cause of Earth's warming trend, with the increase in greenhouse gases coming mostly from the burning of fossil fuels. These fossil fuels, such as oil, coal, and natural gas, are used for many things, including generating electricity, heating buildings, cooking, and running machinery. This results in a large amount of carbon dioxide being put into the atmosphere by human activities.

Scientists also theorize that deforestation, or the cutting down or burning of large areas of trees, has increased the amount of carbon dioxide in the atmosphere because there are fewer trees to change carbon dioxide to oxygen through photosynthesis, and the burning itself releases carbon dioxide. Many countries are now working on ways to reduce the burning of fossil fuels and slow down deforestation.

Deforestation Courtesy of NASA/LBO-ECO Project

Scientists think there are a variety of factors involved in creating climate change. Because the interaction of Earth's spheres is so complex, we still have much to learn about the causes of climate change.

8.7 Summary

- The atmosphere is a thin layer of gases that surrounds the Earth.
- The Earth's atmosphere is divided into five layers: the troposphere, stratosphere, mesosphere, thermosphere, and exosphere.
- The most abundant gas in the atmosphere is nitrogen, with oxygen being the next most abundant.
- Atmospheric pressure is the force exerted on us by the weight of air molecules in the Earth's atmosphere.
- The greenhouse effect is the trapping of energy from the Sun by the Earth's atmosphere. This process keeps the Earth warm.

8.8 Some Things to Think About

- When do you notice the atmosphere the most? What do you notice about it?
- What facts about the atmosphere did you learn that you didn't know before? What was surprising to you?
- Briefly describe each layer of the atmosphere.
- What factors do you think contribute to the formation of layers in the atmosphere?
- How are the different layers of the atmosphere important to life?
- If you live at sea level and then visit a mountaintop that is at an elevation of 3,000 meters (about 10,000 feet), do you think you will notice a difference in the atmosphere? If so, what will you notice and why?
- What challenges would the Moon's lack of atmosphere and low gravity pose for a Moon colony?

- How might some space projects benefit from the Moon's low gravity?
- Describe briefly how the greenhouse effect works.
- Do you think the reduction of the use of fossil fuels would have an impact on the greenhouse effect? Why or why not?

Chapter 9 The Hydrosphere

9.1 Introduction

If you look at a photograph of Earth taken from the Moon, you might notice that our planet looks blue. The blue color of our planet comes from the abundant water found on Earth's surface and in our atmosphere. Life could not exist without this water.

The hydrosphere
Courtesy of NASA, http://visibleearth.nasa.gov

Earth's water is found in oceans, streams, lakes, glaciers, soil, groundwater, and air. All of this water considered as a whole is called the hydrosphere. The term hydrosphere comes from the Greek word *hydro* which means "water." Hydrosphere means "sphere of water." The hydrosphere includes all the water surrounding the Earth. Scientists who study water and the hydrosphere are called hydrologists.

On Earth, water is found in all physical states of matter — solid (ice), liquid, and gas (water vapor). All forms of life on Earth need to have water in order to exist. Some organisms contain up to 95% water, while almost all are more than half water. Human bodies are made up of more than 60% water, and humans cannot survive more than a few days without it.

9.2 The Hydrologic Cycle

Water is constantly moving through various water systems on Earth, going from oceans to the atmosphere, falling as rain and snow on land, and then running back to the ocean in streams and rivers. This continuous movement of water is called the hydrologic cycle, or water cycle. Recall that a cycle is a series of events that repeats itself. In the hydrologic cycle the Sun heats the water in the Earth's oceans, lakes, and streams, which causes some of the water to evaporate. During evaporation, water turns into its gaseous state, or water vapor, and the water vapor escapes into the atmosphere. Once in the atmosphere, the water can return to Earth's surface as precipitation—rain, snow, sleet, and hail.

Another way for water vapor to enter the atmosphere is through sublimation. During sublimation, ice and snow go directly from the solid state of water to the gaseous state without first passing through the liquid state. If you hang out your wash on a day that is below freezing and the air is very dry, you can observe that the water in the clothes freezes to become ice. After a time the ice will have sublimated, leaving your clothes dry even though the temperature is still below freezing. Sublimation also explains why ice cubes left in a freezer over a long period of time gradually become smaller and smaller and finally disappear.

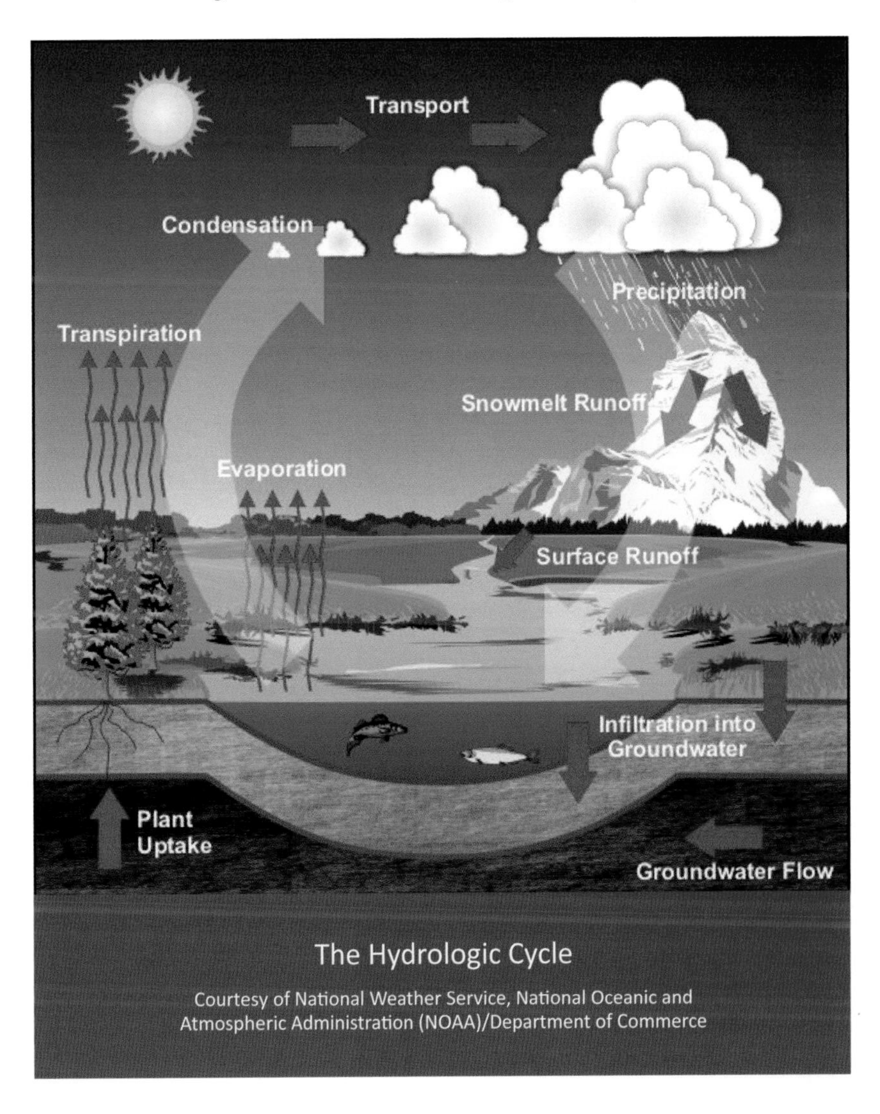

The Hydrologic Cycle
Courtesy of National Weather Service, National Oceanic and Atmospheric Administration (NOAA)/Department of Commerce

The water vapor that has entered the atmosphere condenses (turns into liquid water) to form clouds, and water is returned to the surface of the Earth in the form of rain, snow, and other types of precipitation. Condensation is the process by which water vapor is turned into liquid water.

Some of the precipitation that falls on the land flows from the land into water sources such as lakes, streams, and rivers. Rivers then carry water back to the ocean. Some of the precipitation that falls on land infiltrates, or enters, the ground and flows below the surface of the ground. This water is called groundwater. Eventually groundwater also reaches the rivers and oceans. Water from precipitation is also stored as ice and snow.

You may have noticed that during the hydrologic cycle, water continually changes its form between ice, liquid water, and water vapor, and the cycle of evaporation, precipitation, and water flow keeps repeating over and over. The same water is used over and over, so the water we have today is the same water that existed a very long time ago. This means that dinosaurs drank from the same water that is on Earth today!

9.3 Oceans

Oceans cover roughly 71% of the Earth's surface and hold about 97% of Earth's water. As a result, the oceans play a very important role in the hydrologic cycle.

There are five major oceans in the world. The largest ocean is the Pacific Ocean, the second largest is the Atlantic Ocean, and the third largest is the Indian Ocean. The two smallest oceans are the Southern Ocean and the Arctic Ocean.

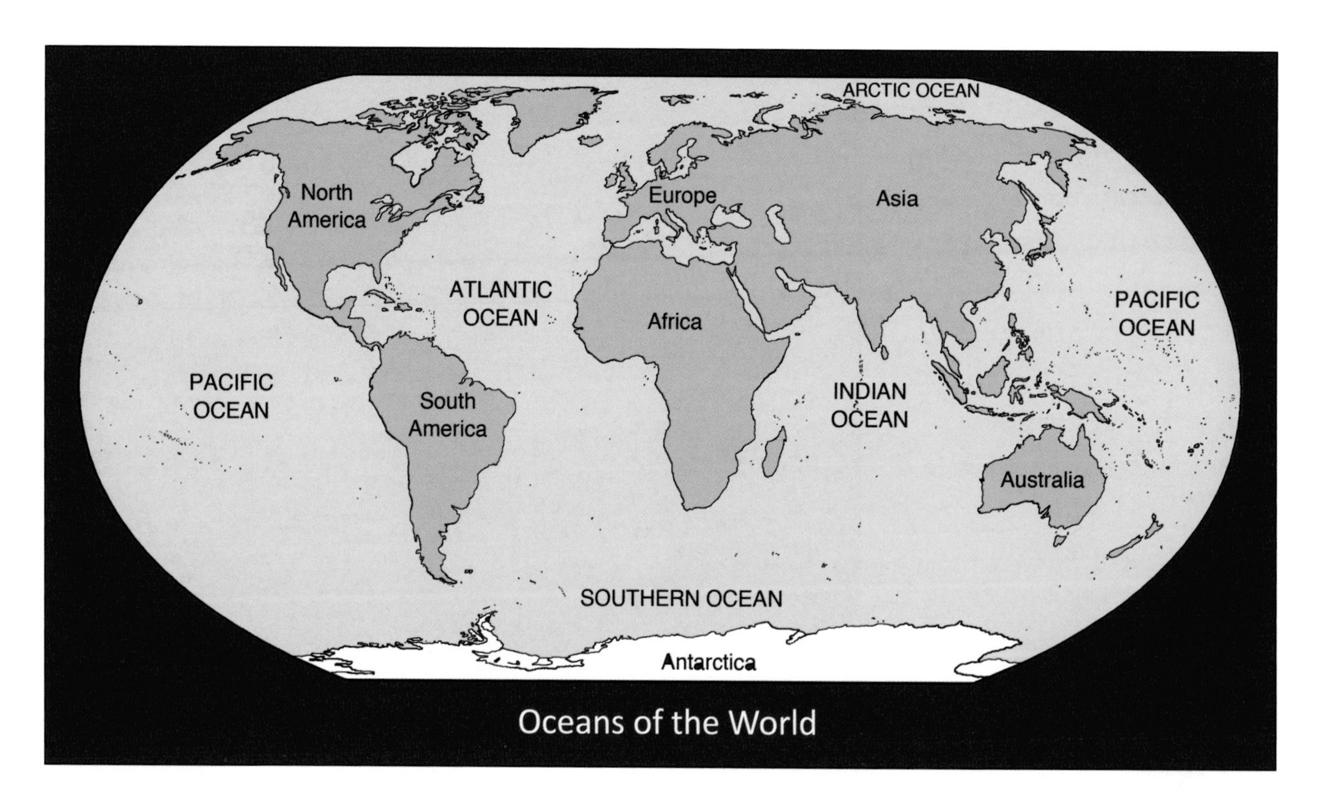

Oceans of the World

Oceans affect Earth's weather and temperature. Oceans absorb incoming heat energy from the Sun. Currents in the oceans then move that absorbed heat energy around the planet. The temperature of an ocean affects the temperature of the air above it, which in turn affects the temperature of the land the air moves across. The hottest waters at the surface of the oceans occur near the equator and can be as high as 36° Celsius (96.8° Fahrenheit). Near the poles the ocean waters at the surface are the coldest with temperatures as low as -2° Celsius (28° Fahrenheit).

Oceans are not the same temperature from top to bottom. There is a warmer upper layer of water that is sun-heated and mixed with oxygen. The deeper part of the ocean forms a colder layer. A thermocline is a thin transition layer found between the upper mixed layer of the ocean and the lower deep water layer. The term thermocline comes from the Greek words *therme* which means "hot" and *kline* which means "bed." A thermocline is the "bed" or layer within which the water temperature changes quickly from the warmer upper layer temperature to the colder deep ocean temperature. Most of Earth's ocean water is found below the thermocline, with only about 5% in the upper mixed layer.

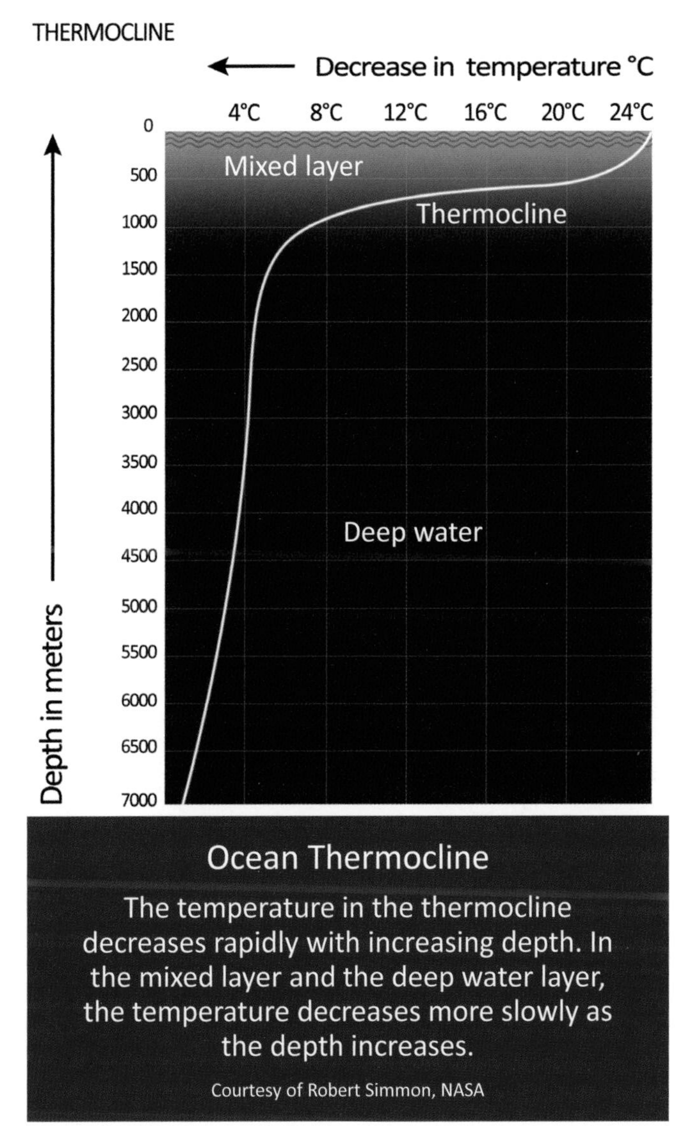

Ocean Thermocline

The temperature in the thermocline decreases rapidly with increasing depth. In the mixed layer and the deep water layer, the temperature decreases more slowly as the depth increases.

Courtesy of Robert Simmon, NASA

Unlike rain water, ocean water contains a mixture of salts and minerals. The amount of salt in ocean water is called its salinity.

Rocks on land are the main source of the oceans' salinity. Rocks are eroded by the action of rain, wind, and other forces. Salts and minerals from rock particles are then dissolved in water from precipitation, and rivers carry the dissolved salts and minerals to the oceans. Salts and minerals also come from the oceanic crust when seawater seeps into the crust and later returns to the ocean, carrying dissolved materials with it. Underwater volcanoes are also an important source of salts when seawater reacts with the hot, erupted rock and dissolves some of the materials it contains.

Ocean water is constantly moving. It travels all the way around the planet by means of currents in the water. Some of these currents are near the surface—in the upper 100 meters (328 feet) of the water — and are caused by the action of wind on the water. Other currents are deep in the ocean. Ocean currents that travel all the way around the world are referred to as the global conveyor belt. Scientists estimate that it takes about 1,000 years for water to travel all the way around the globe.

The global conveyor belt moves warm, upper layer seawater around the globe. This affects temperatures on land, making them more moderate. Also, this circulation of seawater is important to life. Water near the surface becomes depleted of nutrients and carbon dioxide. The deep currents pick up these nutrients and once again bring them to the surface for use by marine plants and animals. The winds on the surface mix oxygen into the water for use by various organisms, and as the oxygenated water sinks, oxygen is mixed into the deeper layers of water where it can be used by organisms living at greater depths.

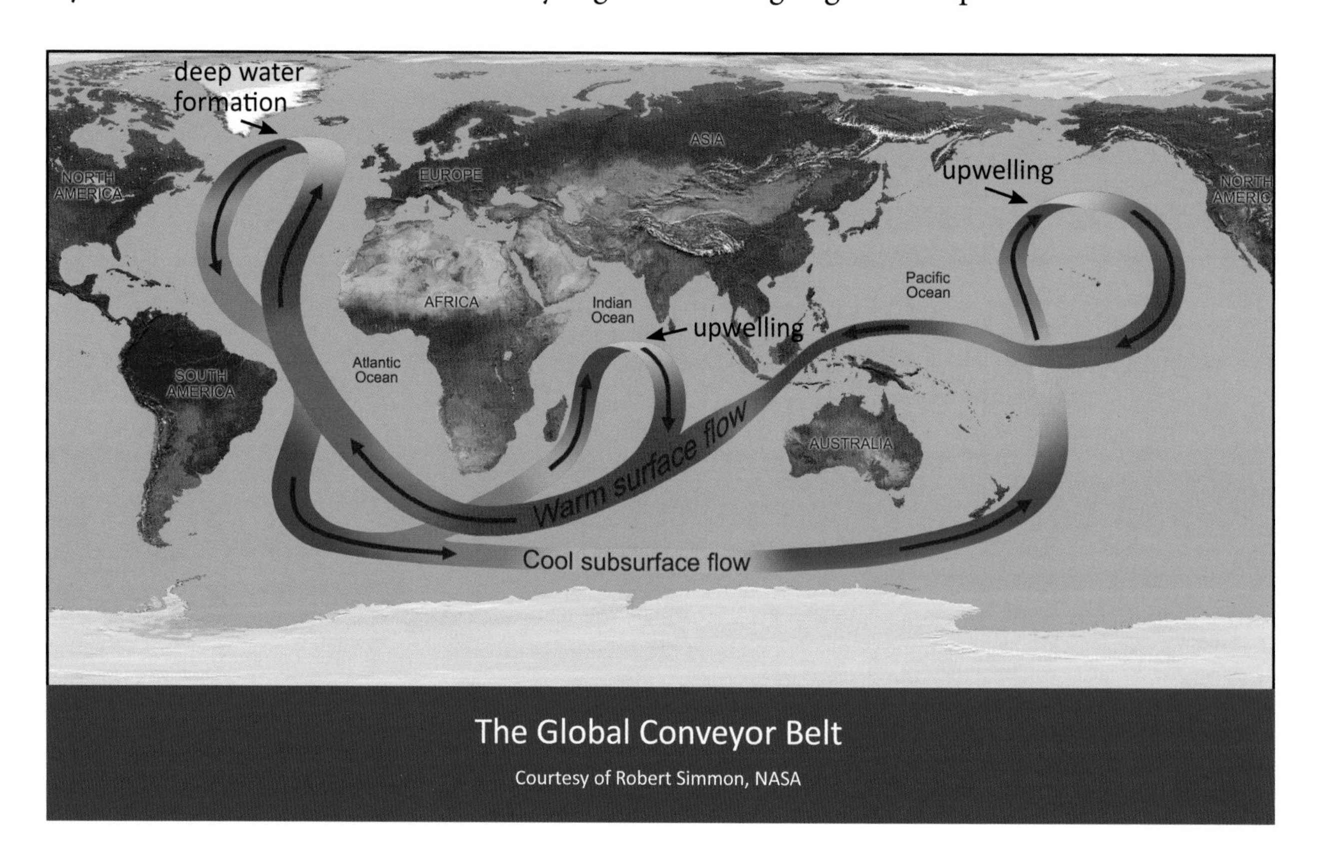

The Global Conveyor Belt

Courtesy of Robert Simmon, NASA

Scientists still have much to learn about the global conveyor belt because it is so large and complex and has many side currents. It is thought that if the global climate becomes warmer, the global conveyor belt currents would slow. The introduction of more fresh water to the oceans from increased rainfall and melting polar ice could cause changes to global ocean currents due to changes in salinity and water temperature. If the movement of warm water from the tropical regions were slowed, it could cause drastic temperature changes in Europe. Marine life could also be drastically affected by changes in water temperature and salinity.

Sea levels are now rising as the water in the oceans becomes warmer, causing it to expand. The melting of polar ice is also contributing to sea level rise. Some coastal areas and small low-lying islands are being impacted by this rise in sea level.

In addition to the global conveyor belt, there are other types of ocean water movement, including waves, tides, and tsunamis. On the surface of the ocean, winds cause waves that can be small and gentle or big and surging. Tides are the periodic rise and fall of oceans caused, in part, by gravitational forces exerted by the Moon. A tsunami refers to an immense wave that can occur when an underwater earthquake, volcanic eruption, or landslide causes a large quantity of water to be suddenly moved. Tsunamis can cause severe destruction in coastal areas.

9.4 Surface Water

Surface water is the water found on land in lakes, rivers, streams, swamps, marshes, and other smaller bodies of water. Although surface water makes up only about three tenths of one percent (0.3%) of the total water found on Earth, it is essential to the lives of all animals that live on land.

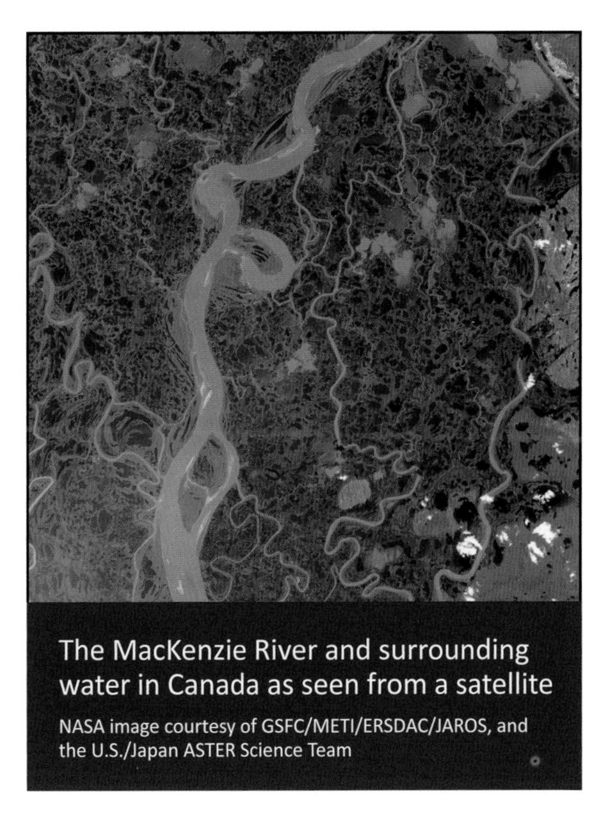

The MacKenzie River and surrounding water in Canada as seen from a satellite

NASA image courtesy of GSFC/METI/ERSDAC/JAROS, and the U.S./Japan ASTER Science Team

Watersheds

Gravity moves water from higher to lower elevations. Precipitation falling on the land flows down slopes into small rivulets that run into larger streams, which in turn flow into rivers. Eventually, the water makes its way to the ocean, feeding lakes, marshes, and swamps along the way. A watershed is the area of land where all the precipitation that falls on it flows, or drains, into a common body of water such as a stream, river, or lake. The watershed of a small stream would cover a small area, while the watershed of a very large river would cover a big area. For example, the Mississippi River in the United States is the fourth largest river in the world with the fourth largest watershed in the world. The watershed for the Mississippi River covers 40% of the area of the lower 48 states, or about 1.2 million square miles, and includes all or part of 31 states and 2 provinces in Canada.

Mississippi River Watershed
Derived from National Park Service illustration

An even larger watershed in the Americas is the Continental Divide, or Great Divide, which is made up of continuous mountain ranges and ridges. On the western side of the Continental Divide, all the water flows toward the Pacific Ocean, and on the eastern side all the water flows into the Atlantic Ocean or the Arctic Ocean. The Continental Divide begins in northwestern Alaska and runs along the Rocky Mountains through Alaska, western Canada, and the western United States. It continues on through Mexico, Central America, and all the way through the Andes Mountains of South America.

An understanding of watersheds is essential when planning activities such as urban development, mining, and agriculture. How does the water flow through the landscape? Where are the high and low areas? Will rivers and streams need to be diverted? If so, what will the consequences be? To keep our water clean, we need to develop solutions to problems such as what to do with industrial pollutants to keep them out of the watershed and how to best deal with storm runoff and waste water in urban settings. Fertilizers and pesticides from agriculture can also cause problems when they enter the watershed.

Lakes, Marshes, and Swamps

A lake is a large body of water surrounded by land. A lake is formed when surface water from precipitation and other flowing water runs into a low area faster than it can flow out. Lakes contain about 87% of all the surface water found on Earth. (This does not include the water in the oceans.)

Humans use the water in lakes for drinking and irrigation and for recreational purposes, such as swimming, boating, and fishing. Many animals depend on lakes for drinking water or as an environment to live in. Plants also live around and in lakes. Because lakes are such an important source of fresh water for humans, animals, and plants, it is important that we keep them clean.

Marshes and swamps are wetlands similar to lakes in that they are surrounded by land, but they are generally shallower than lakes. Marshes usually have grasses and reeds growing in them and often have areas of open water. Swamps most often have trees growing in them. Both marshes and swamps are rich with life and are home to many kinds of frogs, turtles, fish, and birds.

A marsh

Glaciers

A glacier is a large body of ice and snow that moves very slowly down sloped land. In very cold climates, as more snow falls than melts, layers of snow accumulate. Over time, pressure from the weight of the upper layers of snow causes the lower layers to turn to ice. As more and more snow accumulates and turns to ice, the enormous weight of the ice causes it to begin to flow downhill, creating a glacier. As the glacier travels down the slope, its weight crushes some of the rock in the land beneath it, and the bottom of the glacier picks up pieces of rock that get pushed and pulled along. Because it is so heavy and has rocks and debris on the bottom, over a very long period of time the movement of a glacier can carve out deep valleys and other landforms.

The Aletsch Glacier (Aletschgletscher), Valais, Switzerland
The biggest glacier in the Alps at about 23 km (14 mi) long with an area of about 81.7 km2 (31.5 sq mi)
Courtesy of Tobias Alt (Tobi 87), Wikimedia Commons License CC BY SA 4.0

Glaciers, the ice sheets of Antarctica and Greenland, and snow caps on mountains together cover about 10% of Earth's land and are thought to store as much as 70- 75% of all fresh water. It is thought that if all of this ice were to melt, the ocean levels would rise by 70 meters (230 feet).

About 2% of all of Earth's water is stored in glaciers, and about 97% of the world's water is located in the oceans. The remaining 1% or so of Earth's water exists as liquid surface water and groundwater, which are used by human beings, other animals, and plants.

9.5 Groundwater

Groundwater is water that is found in rocks and soil that are below the surface of the land.

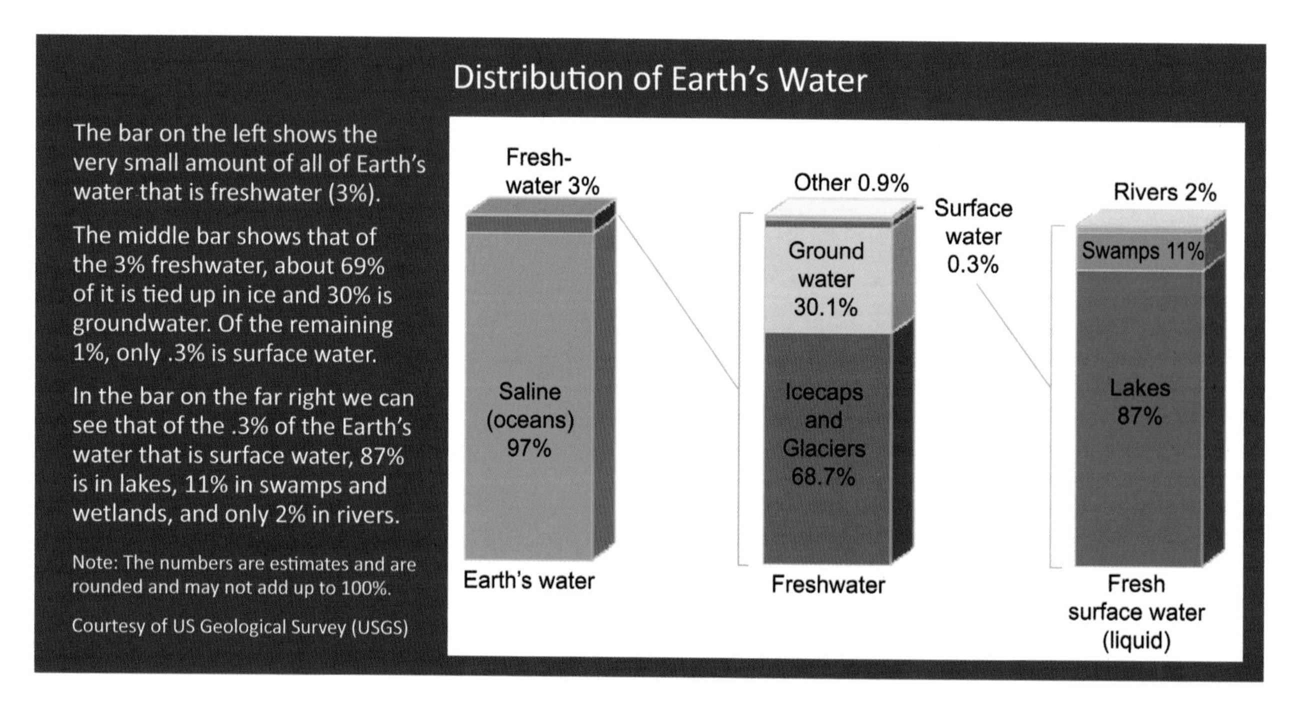

Porous rock

Courtesy of US Geological Survey (USGS), South Florida Information Access (SOFIA)

Groundwater is formed when surface water (such as water in lakes and rivers) or water from precipitation moves downward from the surface, infiltrating the ground and filling the empty spaces between the particles that make up soil and rock. These empty, or void, spaces are called pores, and rock or soil that contains pores is termed porous. Sedimentary rocks tend to be the most porous. Recall from Chapter 3 that detrital sedimentary rocks are made from layers of eroded rock and other debris. The lower layers turn into rock when they are subjected to extreme pressure from the huge weight of upper layers. Because they are formed in this way, sandstones are often porous. On the other hand, igneous rocks, such as granite, are formed from magma deep in the Earth and tend to be very dense and not porous because the particles they are made of are packed too tightly together.

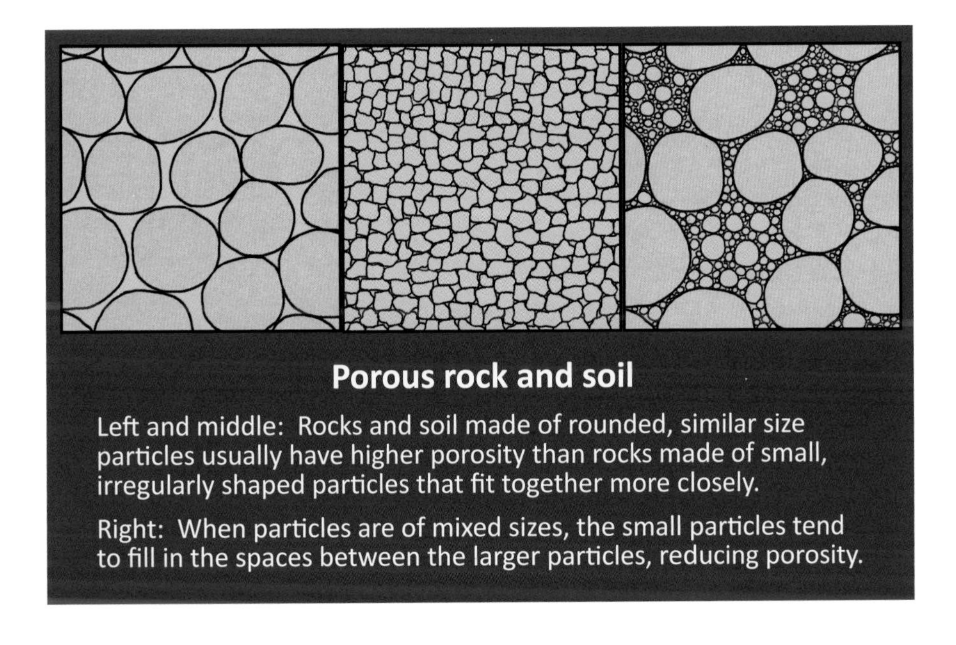

Porous rock and soil

Left and middle: Rocks and soil made of rounded, similar size particles usually have higher porosity than rocks made of small, irregularly shaped particles that fit together more closely.

Right: When particles are of mixed sizes, the small particles tend to fill in the spaces between the larger particles, reducing porosity.

The percentage of the total volume of rock or soil that is made of void spaces is called its porosity. Since porosity is a measure of the size and quantity of the pores in the soil or rock, it also describes the volume of liquid the rock or soil can hold. When the pores are connected and water can flow from pore to pore through the rock or soil, the rock or soil is said to be permeable. It is possible for rock to be porous but not permeable if it has void spaces but they don't connect. On the other hand, rock that is not porous can be permeable if it is highly fractured with cracks that fill with water and are connected. The pores in rock are generally very small, causing the groundwater to flow slowly.

An aquifer is a naturally occurring underground layer of sand, gravel, or rock that has enough porosity and permeability to hold and transmit water. Water that moves downward from the surface may enter an aquifer which can be located anywhere from near the surface to very deep underground. Depending on how the aquifer is structured, it can hold water in one area or transmit water, sometimes over very long distances.

Aquifers can be made of different mixtures of sand and gravel and can also be made of rock. The sedimentary rocks sandstone and limestone generally make the best rock aquifers, although other types of rock can also transmit water. Since sandstone is made of weathered particles of rock and other particles, it can have enough porosity and permeability to make a good aquifer. The greater the porosity and permeability of the rock and the more fractures,

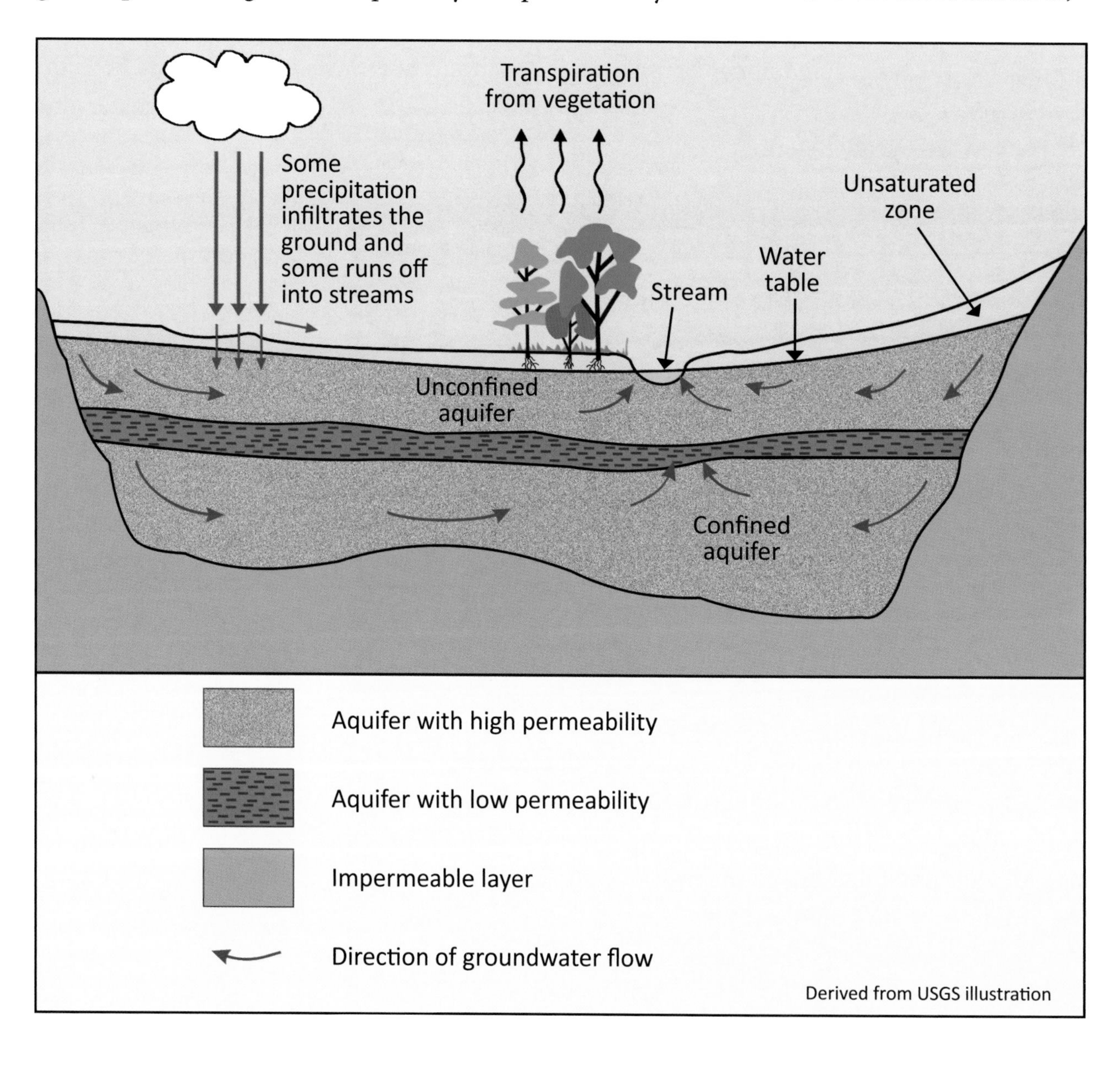

or broken places, it has, the greater the quantity of water that can be stored and the more the water can flow. Sandstone or limestone can be layered with nonporous, less permeable rock, such as shale or siltstone, which acts as a barrier to hold the water within the aquifer. The rock layers in aquifers can be level or tilted and can hold water in one area or allow it to flow long distances.

A limestone aquifer is formed by acidic water that is circulating and dissolves some of the limestone formation over time. This process of dissolution causes openings to form in the rock that range from small tunnels to big caverns. When these openings are well-connected, they can deliver large quantities of water to wells that are drilled into the limestone layer.

A limestone cave

Acidic water has dissolved limestone to form this cave. Minerals have precipitated from dripping water to create chemical sedimentary rock formations.

Courtesy of Pamela Weaver

Because wells can be drilled into aquifers to take water out, aquifers are a vital source of fresh water needed for drinking, washing, bathing, and watering crops and gardens. Some aquifers contain salty water that is not useful for these purposes. In some places a naturally occurring spring can form when part of an aquifer is exposed to the surface, providing a source of water for animals and people. The water in aquifers eventually reaches the oceans as part of the hydrologic cycle.

9.6 Water and Human Interaction

Since the same water is continuously being recycled through Earth's various water systems, the water that was present long ago still exists today. This means that the same water gets used over and over. Therefore, we need to be aware of how people's activities affect the hydrologic cycle so we can understand what we need to do to keep our water clean and available for use.

People pollute water, or make it dirty, when they dump chemicals, garbage, and sewage into rivers, lakes, streams, and oceans. Also, chemicals that are put on the land, such as pesticides, fertilizers, and oil leaks from cars, can enter the groundwater when they get mixed with rainwater and infiltrate the soil. Many people in the world do not have access to clean drinking water, leading to unnecessary illnesses and even death. Water pollution can sicken and kill animals and plants on land and fish, mammals, and plants in lakes and oceans. Research is now being done by different organizations to develop new ways of handling wastes in order to begin to solve these problems. And each of us can contribute to keeping water clean and available by being careful to avoid polluting it and by using less water in dry areas.

Acidic gases are produced by the burning of fossil fuels in power plants, factories, and our houses. Car engines also burn fossil fuels and emit acidic gases as exhaust. These acidic gases are carried up into the atmosphere where they mix with water vapor in the clouds. This causes the rain or snow that falls from these clouds to be acidic, and this acidic precipitation is called acid rain.

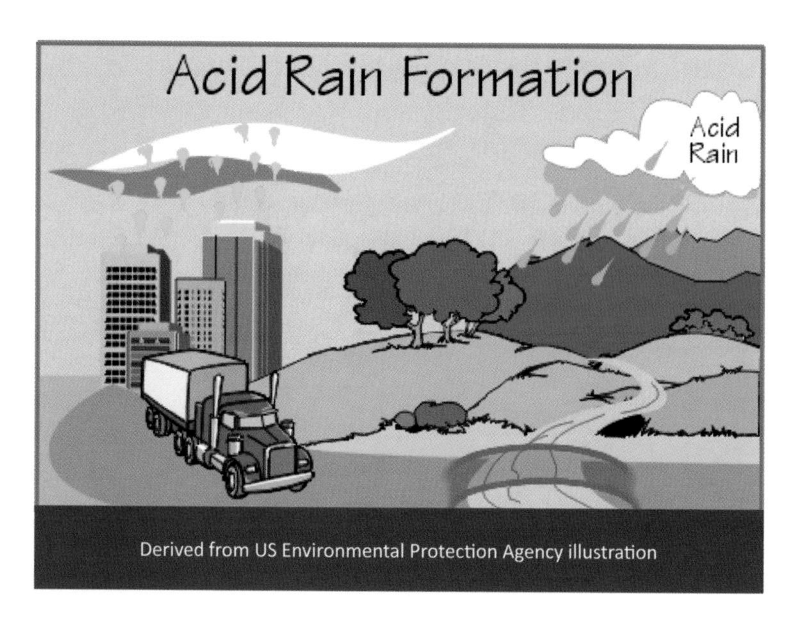

Derived from US Environmental Protection Agency illustration

Effects of acid rain on a forest in Germany

Courtesy of I, ArtMechanic CC BY SA 3.0

Precipitation that is too acidic can pollute lakes and groundwater and kill plants and fish. It can even slowly eat away stone statues! Since clouds can travel long distances before precipitation occurs, acid rain can fall far away from the source of the pollution. It is important that we begin to find ways to curb the release of acidic gases into the atmosphere.

Another problem we're having with our water sources is the over-pumping of water from aquifers. Precipitation that infiltrates into aquifers adds water to, or recharges, them. If more water is pumped out than is recharged, the amount of water in the aquifer begins to decrease. For an aquifer that is recharged slowly, the supply of water it holds can be depleted. If this happens, a formerly productive water well will no longer be able to produce water. Agriculture is a heavy user of water, and in areas experiencing drought conditions, such as in California, the aquifers are being depleted at increasing rates. Depletion of the water in aquifers can also cause land to subside, or sink, when rock no longer has the support of water in its pores. This is happening in some agricultural areas in California and in other places around the world.

Satellites have become an important tool to help scientists study the hydrologic cycle. The data provided is helping scientists gain more understanding of what happens to rain and snow that fall on land—how much of it runs off, how much evaporates, how much is used by plants, and how much water plants put back into the atmosphere through transpiration. Water levels in below-ground aquifers can be measured as can the flow rate of rivers, the height of oceans, ocean wave patterns, changes in the size of ice sheets, moisture in the soil, and even water quality. This data is being used to observe how the hydrologic cycle works and how climate change is affecting the water on our planet. Finding out details about the hydrologic cycle will help scientists find ways to conserve and protect our water.

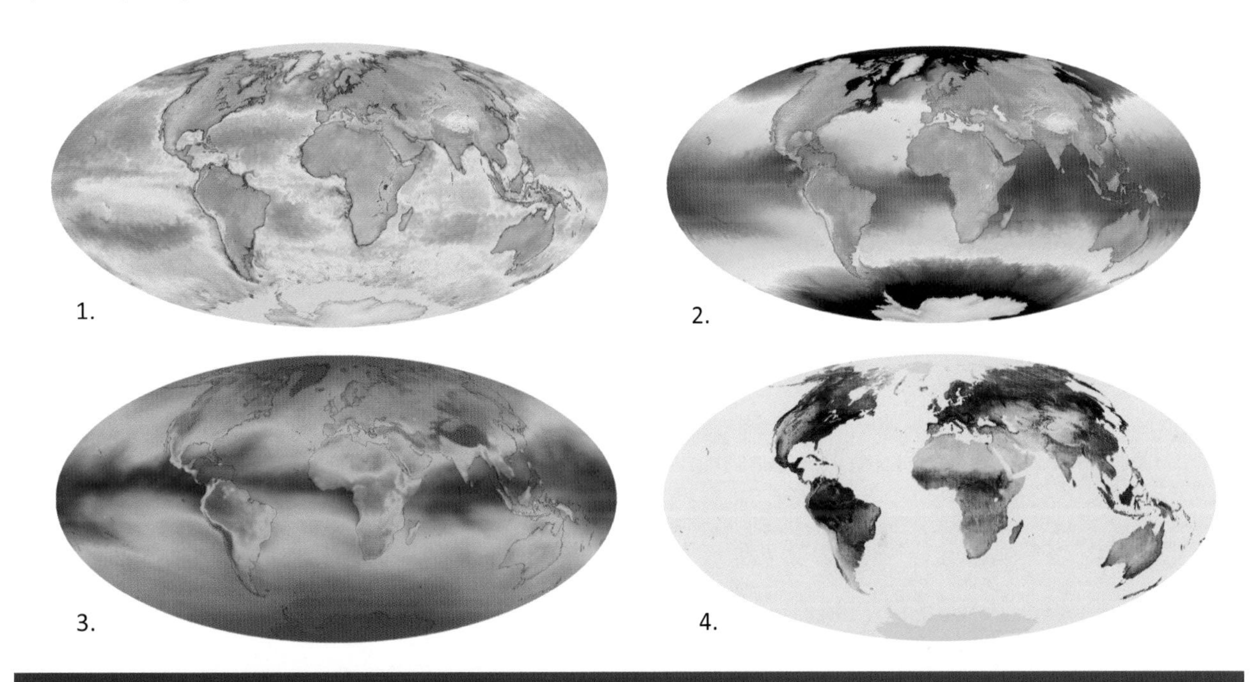

A few types of data collected by satellites

1. Chlorophyll in the oceans; 2. Ocean surface temperatures;
3. Moisture in the atmosphere; 4. Vegetation

Images courtesy of NOAA National Environmental Satellite, Data, and Information Service (NESDIS)

Each of us can help keep our water clean by being aware of how our actions affect the water supply and finding out whether the products we use contribute to pollution. Disposing of trash properly and picking up trash we find on the ground will keep it from entering rivers and oceans. Plastic bags are dangerous to wildlife on land when they eat or get caught in the bags and are dangerous to life in the oceans as well. With a little planning we can greatly reduce our reliance on plastic bags. Reusable shopping bags and reusable glass storage containers can be a good alternative to plastics, and reusable water bottles save the water that goes into producing disposable plastic water bottles and also reduce trash. Since it has been discovered that chemicals in certain sunscreens can damage and even kill coral reefs, we can choose these products carefully to avoid the ones that contain harmful chemicals. We can choose foods that are grown using the fewest pollutants. Household products that contain the fewest and least harmful chemicals can be the ones chosen. Also, there are many ways to reduce water use at home and conserve water. With a little thought and research, you will be able to come up with many ideas and solutions to water pollution and depletion problems. You may be the one who comes up with a great new idea for preserving our water!

Organisms that live on land need fresh water in order to survive, and organisms that live in both fresh water and salt water need clean water if they are to live. We have much work to do to solve the problem of how to keep our water supply clean and abundant.

9.7 Summary

- Hydrosphere is the term for all the water that surrounds the Earth.
- The hydrologic cycle is the process by which water continuously moves around the Earth.
- About 97% of all water on Earth is found in the oceans.
- Ocean salinity is caused mostly by erosion of rocks on land.
- Surface water is found in lakes, marshes, swamps, streams, rivers, and glaciers.
- A watershed is the area of land where all the precipitation that falls on it flows, or drains, into a common body of water.
- Groundwater is water that is found in the soil and rocks below the surface of the Earth.

- An aquifer is a naturally occurring underground layer of sand, gravel, or rock that has enough porosity and permeability to hold and transmit water.
- Humans and human activities contribute to pollution of air and water.

9.8 Some Things to Think About

- How many different ways can you think of that you have interacted with water?
- Describe how you think the hydrologic cycle works in your area.
  Where is water found? When does it change states? How does it move? Do the seasons affect the hydrologic cycle? What else can you observe?
- Why do you think different marine animals can live in a certain part of the ocean but not in other parts?
- What do you think would happen to marine animals if the salinity or temperature changed in the area of the ocean where they live?
- Get a topographical map of your area that shows rivers, lakes, streams, and land elevations. Use arrows to mark the direction you think water flows in the rivers and streams and the direction that rain flows off the land. Draw a line around the area or areas that you think are watersheds.
- Take your field notebook and go outside for a hike. See what you can observe about surface water - where it is and how it moves.
- Have you ever turned on a faucet and wondered where the water comes from? Do some research to find out where the water in your house comes from and how it gets there.
- What groundwater issues do you think you have in your area? Is water plentiful? Scarce? Clean? Polluted? Treated to make it safe to drink?
- When have you noticed water being used unnecessarily or in excess?
- What have you noticed in your area that could be contributing to water pollution?
- What actions can you take to use less water and reduce pollution?

Chapter 10 The Biosphere

10.1 Introduction

The biosphere is the part of Earth that includes all living things and their relationship to each other. The Greek word *bios* means "life," so biosphere means "life sphere." All of the ecosystems and their interactions with the hydrosphere, geosphere, and atmosphere make up the biosphere.

The biosphere includes the part of the atmosphere where birds and insects fly, the part of the geosphere where animals roam and plants and microbes live in the soil, the hydrosphere where fish swim and frogs jump in lakes and rivers, and as far as the depths of the ocean where strange-looking creatures are found. The biosphere interacts with the energy from the Sun and with forces deep within the Earth, such as those that create earthquakes and volcanoes.

Earth's biosphere

Courtesy of NASA/Goddard Space Flight Center Scientific Visualization Studio

Because living things don't just live on the surface of the Earth but interact with it, the biosphere is constantly changing and varies greatly from one geographic location to another. For example, soil is a combination of both rock and organic matter, and soils vary depending on the type of rock they came from and the type of plants and animals that have interacted with them. For instance, rich soils that support abundant plant life contain a large amount of organic matter that comes from dead plants and animals that have decomposed. Rich soils can be found in marshy, wet, and tropical locations and other places where plant and animal life is abundant. On the other hand, sands don't contain abundant organic matter and are found in drier and more arid climates where fewer organisms live.

A beaver dam changes water flow
Courtesy of US Fish and Wildlife Service/by Luther C. Goldman

Although Earth's landforms are largely determined by moving plates, volcanoes, earthquakes, and glaciers, plants and animals also contribute to the formation of Earth's surface features. Plants and animals can contribute to weathering and erosion. For example, plants growing in small cracks in rocks can, over time, create fractures in the rocks, breaking them into smaller pieces with their roots. Also, animals move rocks and soil as they travel around and dig burrows, and erosion can occur as water flows along trails that animals have made.

10.2 Cycles in the Biosphere

Different elements are constantly being reused as they move in cycles through the biosphere and other Earth "spheres." These cycles are called biogeochemical cycles because they involve biological organisms (bio); non-biological components such as water, air, and the lithosphere (geo); and the chemicals being cycled. Each cycle contains an exchange pool where elements and molecules are cycled relatively quickly between living and nonliving components of the cycle, and one or more reservoirs where elements and molecules are stored for longer periods of time. In fact, all the chemical elements that make up any organism are reused during biogeochemical cycles. When an organism dies, it is broken down into elements and molecules that can be used to make new organisms, or stored and used later. The same elements are used over and over. Some biogeochemical cycles that are important to the biosphere are the oxygen cycle, the carbon cycle, the nitrogen cycle, and the energy cycle. In the next few sections we will take a closer look at these cycles.

The Oxygen Cycle

The exchange pool of the oxygen cycle mainly involves the interaction of photosynthetic organisms and animals and occurs in the biosphere, atmosphere, and hydrosphere. During photosynthesis, plants and microscopic aquatic organisms use sunlight, water, and carbon dioxide to produce sugars for food, giving off oxygen as a waste product. Animals then breathe in this oxygen and breathe out carbon dioxide as a waste product. The cycle continues when plants and phytoplankton use the carbon dioxide produced by animals. For oxygen to be usable by animals, it has to be "free" oxygen (not combined with other elements) in the molecular form of O_2.

Like land animals, aquatic animals need oxygen to live. They can't use the oxygen that is bound in water molecules but instead rely on O_2 molecules that are dissolved in water. Free oxygen enters the water from the atmosphere and also from photosynthesis by phytoplankton. Phytoplankton is the name for a group of microscopic organisms that live in both fresh and salt water and includes bacteria, protists, and single-celled plants. It is estimated that phytoplankton are responsible for 50% or more of the total oxygen produced by photosynthesis. The word phytoplankton comes from the Greek words *phyton* which means "plant" and *plankton* which means "wandering, drifting." These

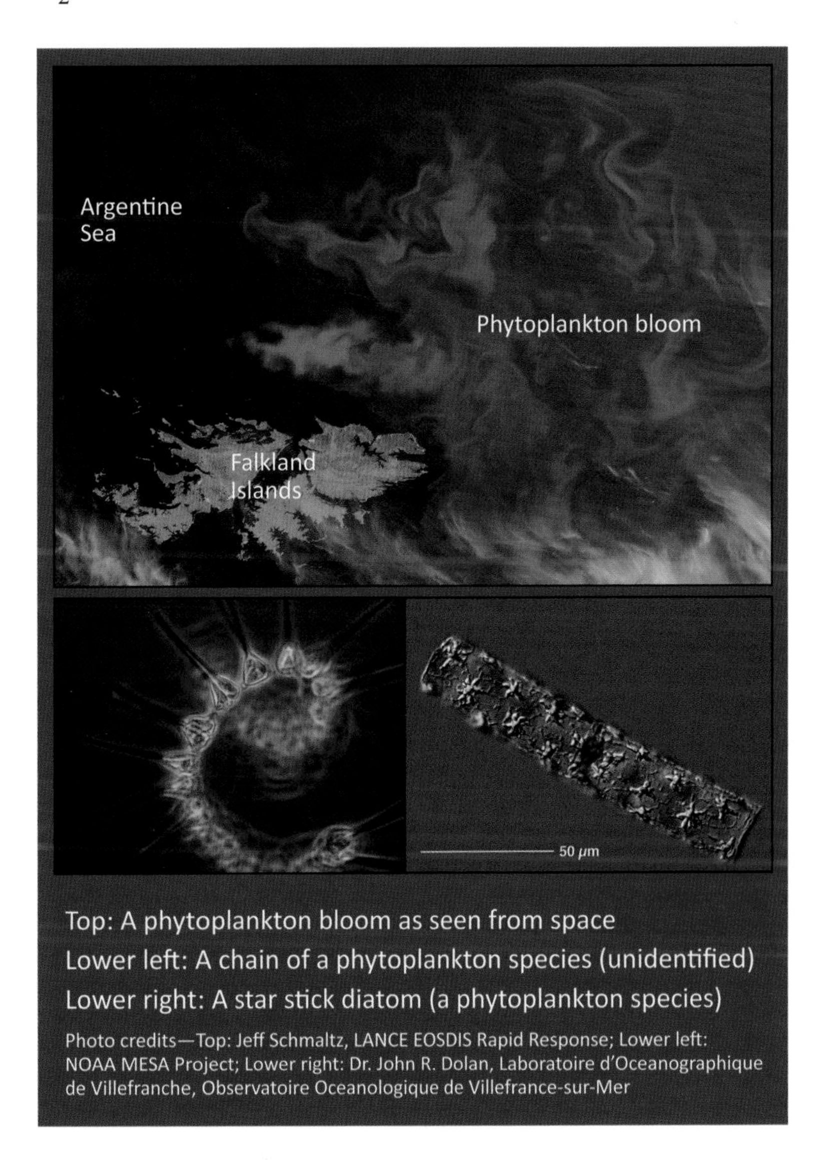

Top: A phytoplankton bloom as seen from space
Lower left: A chain of a phytoplankton species (unidentified)
Lower right: A star stick diatom (a phytoplankton species)
Photo credits—Top: Jeff Schmaltz, LANCE EOSDIS Rapid Response; Lower left: NOAA MESA Project; Lower right: Dr. John R. Dolan, Laboratoire d'Oceanographique de Villefranche, Observatoire Oceanologique de Villefrance-sur-Mer

tiny organisms float freely through water, using carbon dioxide for photosynthesis and producing oxygen that goes into the water and the atmosphere. Phytoplankton are an important food source for many aquatic animals.

In addition to the oxygen added to water by phytoplankton, winds and water turbulence in streams and rivers help mix oxygen from the atmosphere into the water. Dissolved oxygen levels in water vary with temperature and salinity, with warm water holding less oxygen than cold water, and salt water holding less oxygen than fresh water. Another factor that determines how much dissolved oxygen is in a body of water is the amount of decomposition of organic matter that is occurring. Decomposition of matter from dead animals and plants uses oxygen, so if there is too much decaying matter in a body of water, the oxygen can be reduced or depleted, causing organisms to sicken or die.

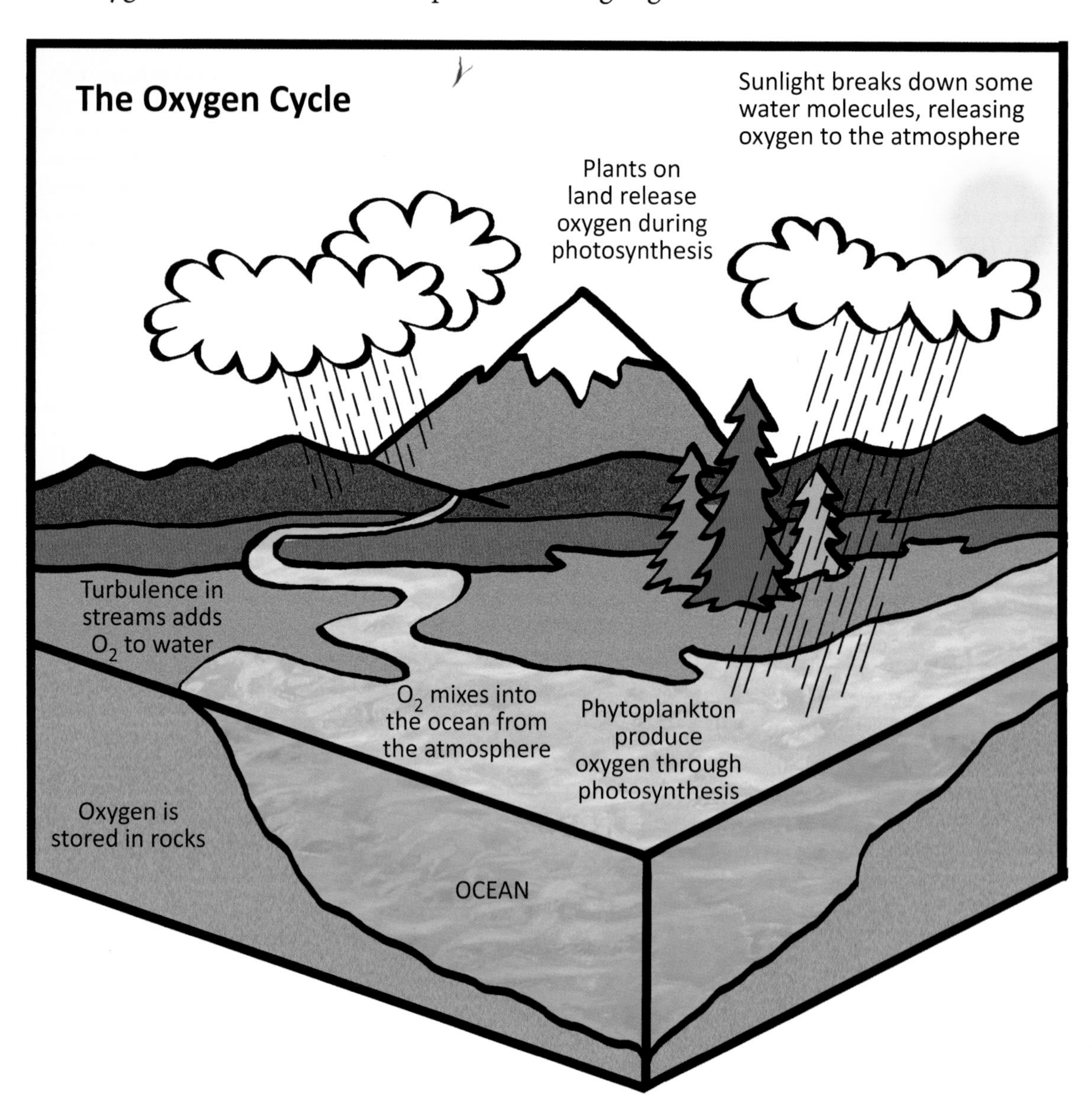

The amount of dissolved oxygen in water helps determine which organisms are able to live in a particular location. For example, crabs, worms, and bacteria can live at the bottom of bodies of water where there is less dissolved oxygen. Fish that are adapted to live in shallow water need more dissolved oxygen. Aquatic plants produce oxygen when there is sunlight, but at night they use dissolved oxygen from the water during cellular respiration when they are converting sugars produced during photosynthesis into energy. Cellular respiration occurs in both plants and animals. During cellular respiration carbon dioxide is given off as a by-product of the chemical reactions that turn nutrients into energy.

The main oxygen reservoir is the lithosphere where oxygen is found in chemical compounds that make up rocks. In fact, it is estimated that 99.5% of all the oxygen on Earth is stored in rocks. Chemical processes that break down rocks can release free oxygen into the atmosphere, while other processes (such as rusting) use oxygen to form new molecules. The second oxygen reservoir is the atmosphere. In addition to photosynthesis, some oxygen is added to the atmosphere as a result of lightning, which causes chemical reactions in the air that produce ozone (O_3). Also, high energy ultraviolet light entering the atmosphere can break apart oxygen-containing molecules in the air, releasing oxygen. The third oxygen reservoir is the biosphere and includes all the oxygen in living organisms.

The Carbon Cycle

Carbon is an element that is needed by all living things, making the carbon cycle vital to the existence of life. With the exception of water molecules, carbon is found in just about all the molecules that make up the human body. About 18% of the human body is carbon.

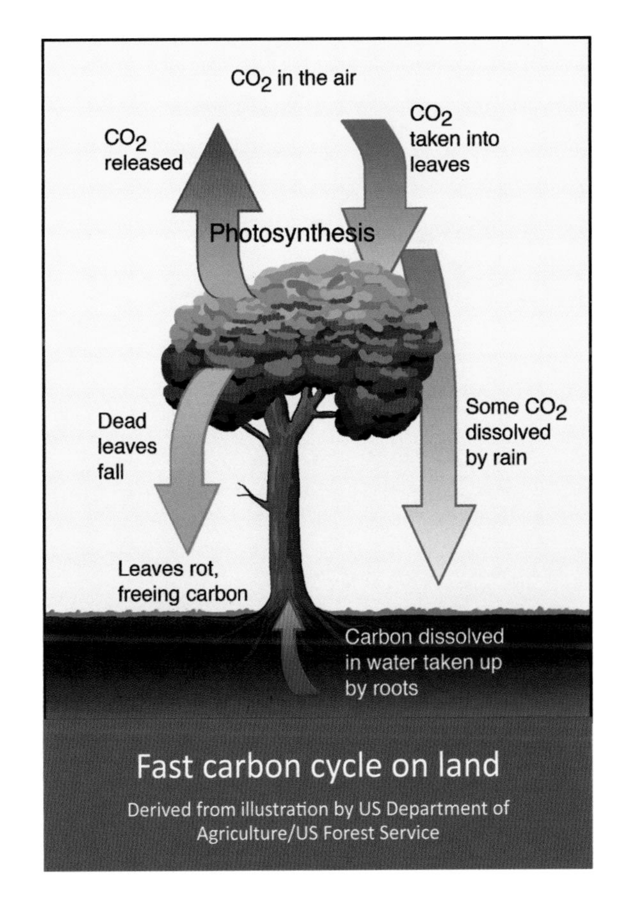

Fast carbon cycle on land
Derived from illustration by US Department of Agriculture/US Forest Service

We can think of the carbon cycle as having two different parts—the fast carbon cycle and the slow carbon cycle. The exchange pool where carbon is cycled more quickly between living and nonliving components is found in the fast carbon cycle and involves the biosphere, atmosphere, lithosphere, and hydrosphere.

In the fast carbon cycle, photosynthesis once again plays an important role. Plants and phytoplankton in the biosphere use carbon dioxide as a source of carbon and combine the carbon with other elements to form the organic molecules necessary for life to exist. When animals eat plants or other animals that have eaten plants, carbon is transferred to the animals' bodies. Carbon dioxide is released into the atmosphere as organic matter decays, as cellular respiration takes place in plants, and as animals breathe.

The oceans play a large part in the carbon cycle and are the largest carbon reservoir after rocks. Recall from Chapter 9 that oceans have an upper mixed layer and a lower deep water layer. In a similar way to oxygen, winds mix atmospheric carbon dioxide into the upper level of the oceans, and carbon dioxide is also released from the upper level of the oceans into the atmosphere. The amount of carbon dioxide taken into the oceans and the amount released into the atmosphere ideally are in balance. Carbon enters the lower layer of the oceans when organisms die and the organic matter sinks. Rocks on the ocean floor dissolve, releasing carbon into the water, and underwater volcanic eruptions add some carbon to the oceans. Recall from Chapter 9 that water in the upper and lower layers of the ocean gets mixed, moving carbon from one layer to the other. The circulation of water upward from the lower layer also brings with it important nutrients to organisms that live closer to the surface.

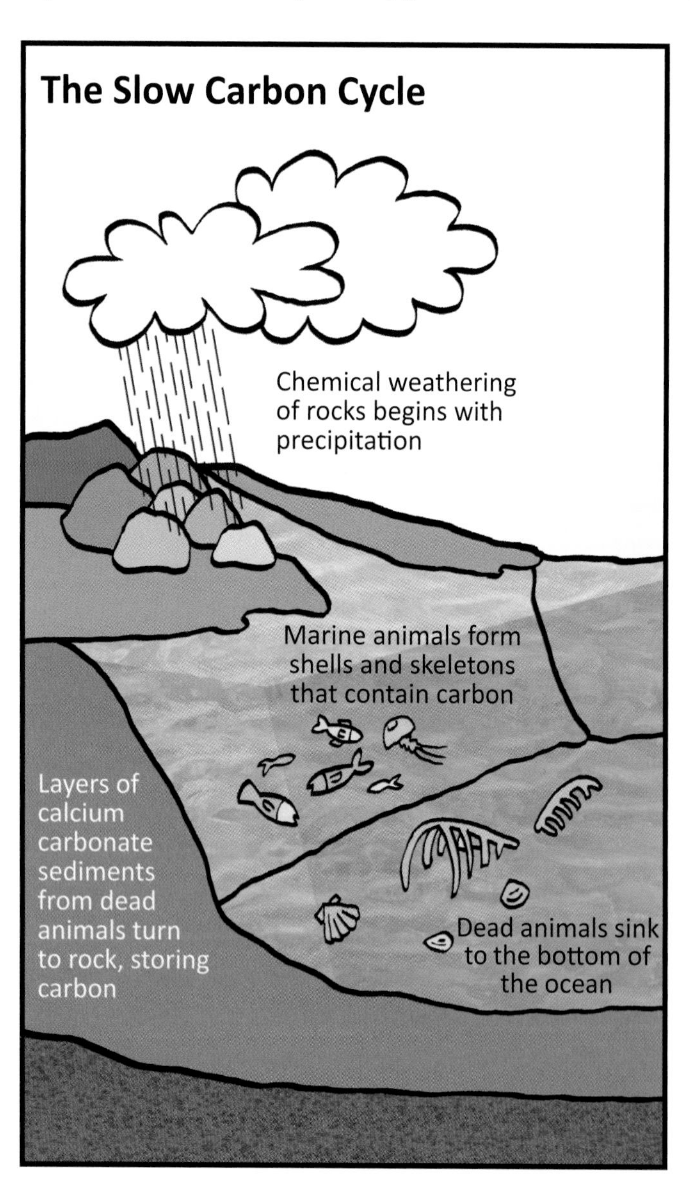

The slow carbon cycle involves rocks. Like oxygen, the main reservoir of carbon is rocks in the lithosphere. Carbon-containing rocks are generally formed under the ocean.

The Nitrogen Cycle

The nitrogen cycle is another cycle that is important to the biosphere. Nitrogen makes up about 78% of Earth's atmosphere and is essential for all life as it is an important component in proteins, DNA, and other molecules. In the exchange pool, plants get nitrogen from the soil, and animals get nitrogen by eating plants or other animals that have eaten plants.

Nitrogen from the air enters the soil as a gas molecule (N_2). Nitrogen in this form is not usable by plants and must undergo a process called nitrogen fixation during which bacteria in the soil convert the nitrogen to a usable molecule. This process is also referred to as "fixing" the nitrogen. Some bacteria that fix nitrogen are free-living and exist in the soil without being associated with any particular plant. Other bacteria have a symbiotic (or interdependent) relationship with plants. These bacteria live in nodes on the roots

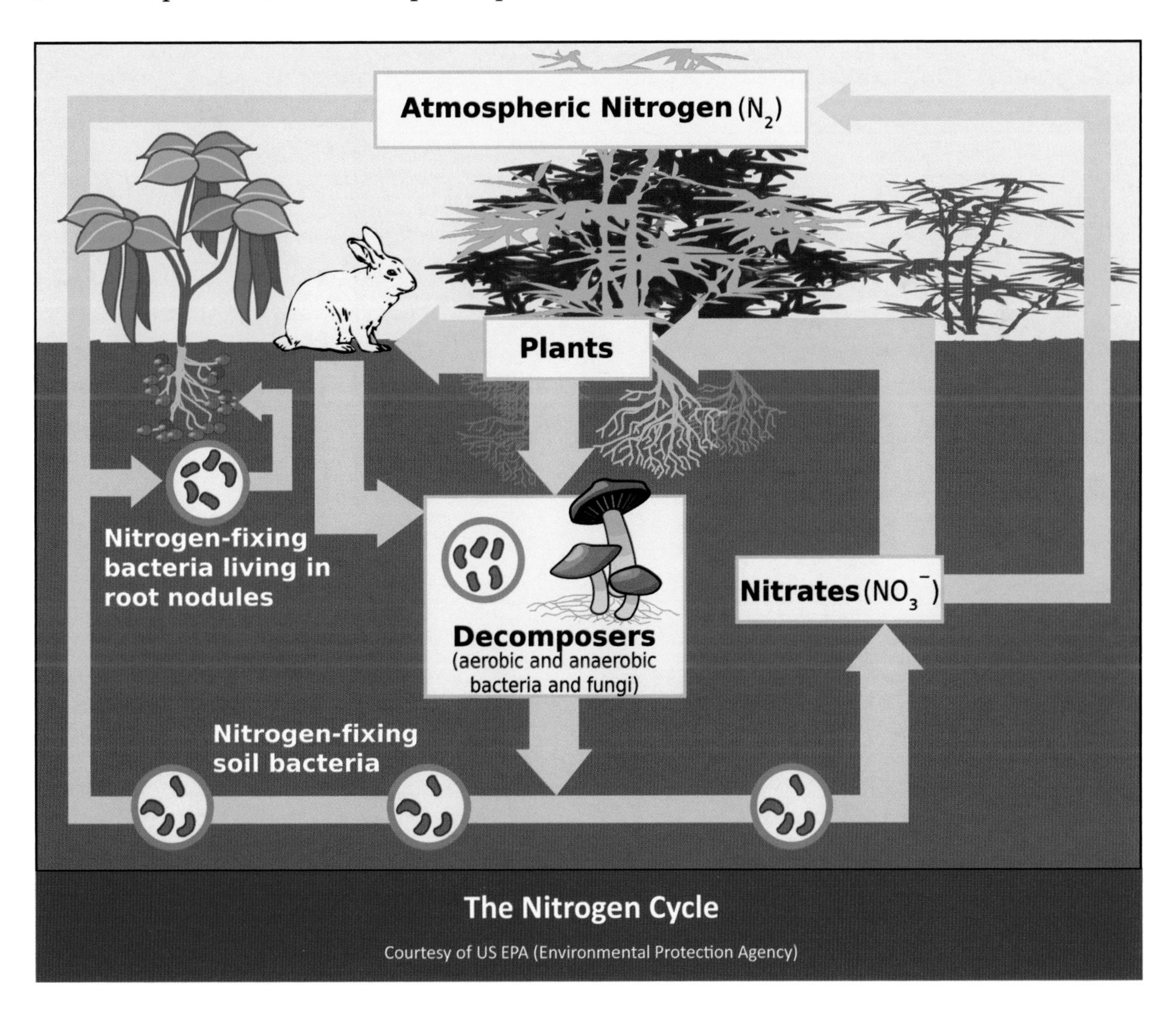

The Nitrogen Cycle

Courtesy of US EPA (Environmental Protection Agency)

where they fix nitrogen that can be used by the plants, and in exchange the bacteria use carbohydrates from the plants.

In the oceans and other bodies of water, atmospheric nitrogen (N_2) is dissolved in the water, and the nitrogen cycle goes through similar steps beginning with those phytoplankton that are able to fix nitrogen.

The Energy Cycle

The energy cycle begins with heat and light energy from the Sun. Although most of the heat and light that reaches Earth is reflected back into space, some of it is captured by plants to make carbohydrates.

Plants use the Sun's energy from the visible portion of the electromagnetic spectrum to make sugars via photosynthesis. Animals get this energy from the Sun indirectly by eating plants or other animals that have eaten plants.

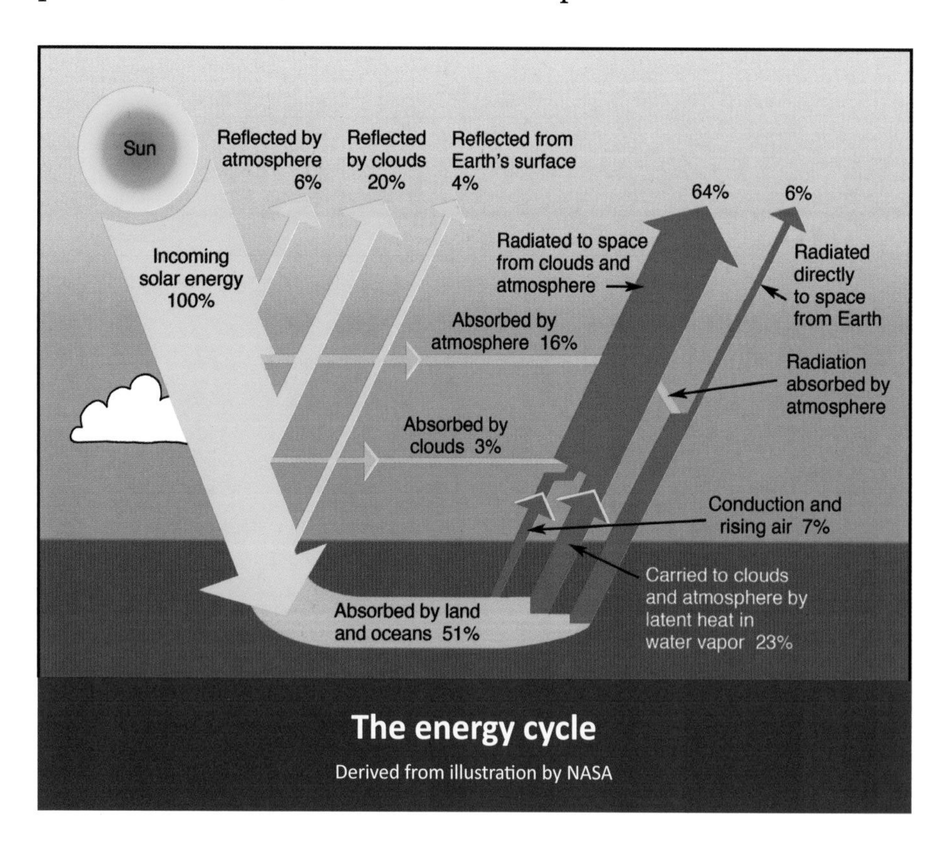

The energy cycle
Derived from illustration by NASA

The atmosphere plays a large role in how energy from the Sun enters the biosphere. The ozone layer absorbs most but not all of the ultraviolet radiation from the Sun, protecting the biosphere from harmful rays. Ultraviolet radiation has shorter, more energetic wavelengths than visible light and can damage living cells. However, humans need some skin exposure to the ultraviolet B part of the electromagnetic spectrum in order for the body to produce the vitamin D used to make healthy bones and other tissues.

Recall that greenhouse gases such as carbon dioxide, water vapor, and methane are found in the atmosphere and cause the greenhouse effect. These gases absorb some of the heat (infrared radiation) coming from the Sun during the day. At night greenhouse gases help keep Earth warm by preventing too much heat from radiating back into space as the Earth cools. When these processes are in balance, the yearly average temperature of the Earth stays about the same from year to year. This is important for plants and animals because different species are adapted to live in certain yearly temperature ranges. For example, some organisms can have difficulty surviving if they experience summers that become too hot and too long, or for others, winters that become too warm and too short.

In Section 10.4 we will see how scientists track energy as it moves through living things in a food web.

10.3 Where Is Home?

Why do plants and animals find homes in certain areas but not in others? Why can't they live just anywhere? In order to be able to live in a certain location, an organism has to have all of its needs met, including enough of the right kind of food, the right amount of water, the right weather, a way to find a mate, and the right kind of place for its young to grow. Some organisms have very specific requirements that greatly limit where they can live, while others are more adaptable and can live in places with a broader range of conditions.

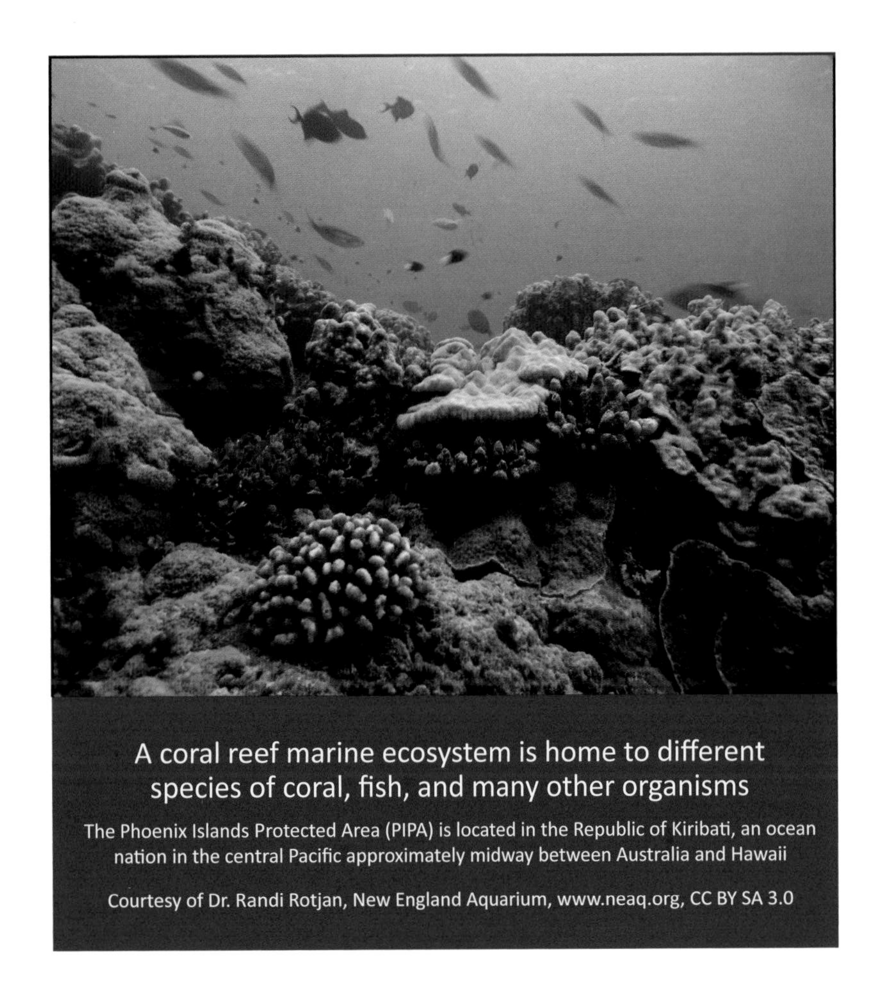

A coral reef marine ecosystem is home to different species of coral, fish, and many other organisms

The Phoenix Islands Protected Area (PIPA) is located in the Republic of Kiribati, an ocean nation in the central Pacific approximately midway between Australia and Hawaii

The study of the interaction and relationships of living things with the world around them is called ecology. The conditions surrounding a living thing are called the environment that it lives in. An environment includes both living and nonliving things, such as plants, animals, weather, soil, and water sources. The Sun, atmosphere, hydrosphere, lithosphere, and magnetosphere are also part of an environment. We can see that what makes a good home for one organism might not be a good home for another organism.

Barn swallow *(Hirundo rustica)* babies in nest

The barn swallow gets its name because it requires a habitat where it can find a protected space, such as an open building structure or under a bridge, to build a nest made of mud lined with feathers and grasses.

As living things interact with their environment, they form an ecosystem, or ecological system. Ecosystem refers to a specific area that contains a community of organisms that exist under similar conditions. An ecosystem can be as large as a tropical rainforest or smaller than a puddle of water and includes both biological organisms and non-biological components, called abiotic factors. The prefix "a-" is from Greek and means "not." Abiotic factors are the non-living chemical and physical aspects of an ecosystem, such as amount of sunlight and moisture, temperature, altitude, and type of soil. Biotic and abiotic factors in an ecosystem are interdependent, and a change in one factor in an ecosystem can make changes occur to other parts of the ecosystem. For example, a lengthy drought can cause plants to die from lack of water. Animals that can't find enough food to eat or enough water to drink can die or fail to reproduce. Fish and other organisms that live in the water will die if the water source dries up. There may be more soil erosion and animals may migrate to other areas, affecting other ecosystems. On the other hand, if an ecosystem experiences changes that are beneficial, plants and animals can thrive. For example, if a desert gets more rainfall than usual, it can lead to an increase in the number and health of plants, giving animals more to eat, which may lead to an increase in animal populations in the ecosystem.

The place in an ecosystem where a particular organism lives is called its habitat. A habitat contains the set of conditions that a specific type of organism needs in order to live and reproduce and can house organisms as small as microbes or as large as whales. An ecosystem can contain many different habitats. All the organisms of the same species that live in the same habitat are called a population, and all the populations of the different habitats in an ecosystem together are called an ecological community.

A tree squirrel and a ground squirrel live in different habitats

Top: Red Squirrel, Courtesy of Bill Thompson, USFWS
Bottom: Columbian Ground Squirrel, Courtesy of The Lilac Breasted Roller, CC BY SA 2.0

While a habitat is a physical location within an ecosystem that has the right conditions for a particular species to live, a niche is the role the species plays within the ecosystem. How does the organism interact with other organisms within the community? How does it interact with resources such as water supply? What is its place in the food chain? Does it make its own food? What organisms does it eat? What organisms eat it? For example, one role of tree squirrels in an ecosystem is to be a spreader of seeds. New trees and berry bushes grow from nuts and seeds that squirrels have buried in the ground in different locations to be eaten later. The nuts and seeds that are forgotten can grow into new trees or bushes. Also, squirrels provide food for predators and can help control some bug populations by eating them.

A biome is a specific type of ecosystem. A biome is a large region that is home to plants and animals that require a similar environment. A biome is defined by the climate, soils, and plant life existing within it. There are watery as well as terrestrial biomes, and the marine water, or ocean, biome is the largest of all. See the chart on the following page for a description of some of Earth's biomes.

Biomes

Photo	Biome
1	**Tundra** a very cold, vast, treeless area found around the North Pole; most plants are low-growing, such as shrubs, grasses, and mosses; the ground below the surface doesn't thaw
2	**Boreal Forest (Taiga)** a cold biome with a forest of mostly conifers (evergreen trees that have cones and needle-like leaves); found in the northern areas of the globe
3	**Temperate Deciduous Forest** warm summers with cold to mild winters; deciduous trees (those that shed their leaves in winter)
4	**Temperate Grassland** hot summers and cold winters; vegetation consists of grasses with few trees or shrubs
5	**Desert** an area that gets very little precipitation; plants are adapted to live with small amounts of water
6	**Savanna** found on either side of the equator; warm temperatures; grassland with shrubs and scattered trees
7	**Tropical Rainforest** warm climate with a large amount of precipitation; located near the equator; vegetation is thick and very diverse with very tall trees
8	**Chaparral** very hot and dry climate with mild winters; plants have adapted by developing leaves that hold moisture
9	**Fresh Water** a water environment that has low salinity, such as lakes and rivers; can contain aquatic plants and algae
10	**Marine Water** oceans; saltwater environments; vegetation includes kelp and sea grasses; algae and phytoplankton are often abundant

Photo Credits: 1. US Fish and Wildlife Service, 2. Mike Przbyla/US Fish and Wildlife Service, 3, 4. National Park Service, 5. US Fish and Wildlife Service, 6. Gary M. Stolz/US Fish and Wildlife Service, 7. Cecil Schwalbe/US Geological Survey, 8. Nicole M. DeCrappeo/US Geological Survey, 9. David Spencer/US Fish and Wildlife Service, 10. Kevin Lafferty/US Geological Survey

In this example, the squirrel's habitat is contained in the larger forest ecosystem which is within the very large temperate deciduous forest biome

10.4 The Food Web

An important part of understanding an ecosystem is learning how different organisms get their food. Each organism can be described as part of a food chain that relates the order in which particular organisms in an ecosystem feed on each other. A food chain is made up of different levels, called trophic levels, that show an organism's position on the food chain. Broadly speaking, there are producers that make their own food by photosynthesis (autotrophs), animals that eat plants (herbivores), animals that eat other animals (carnivores), and decomposers that break down dead organisms.

A food chain

Food chains are useful for describing what food particular organisms eat in a progression from plants through apex predators, but because organisms have more than one source of food and because the organisms in an ecosystem are interrelated, the idea of the food web was developed to show more complicated relationships. A food web is made up of the many different food chains in an ecosystem and shows how an individual species feeds and is fed on by more than one other species. Ecologists use food webs to reveal different patterns, including how species interact directly and indirectly, how energy is transferred between different types of organisms, and how an ecological community is structured.

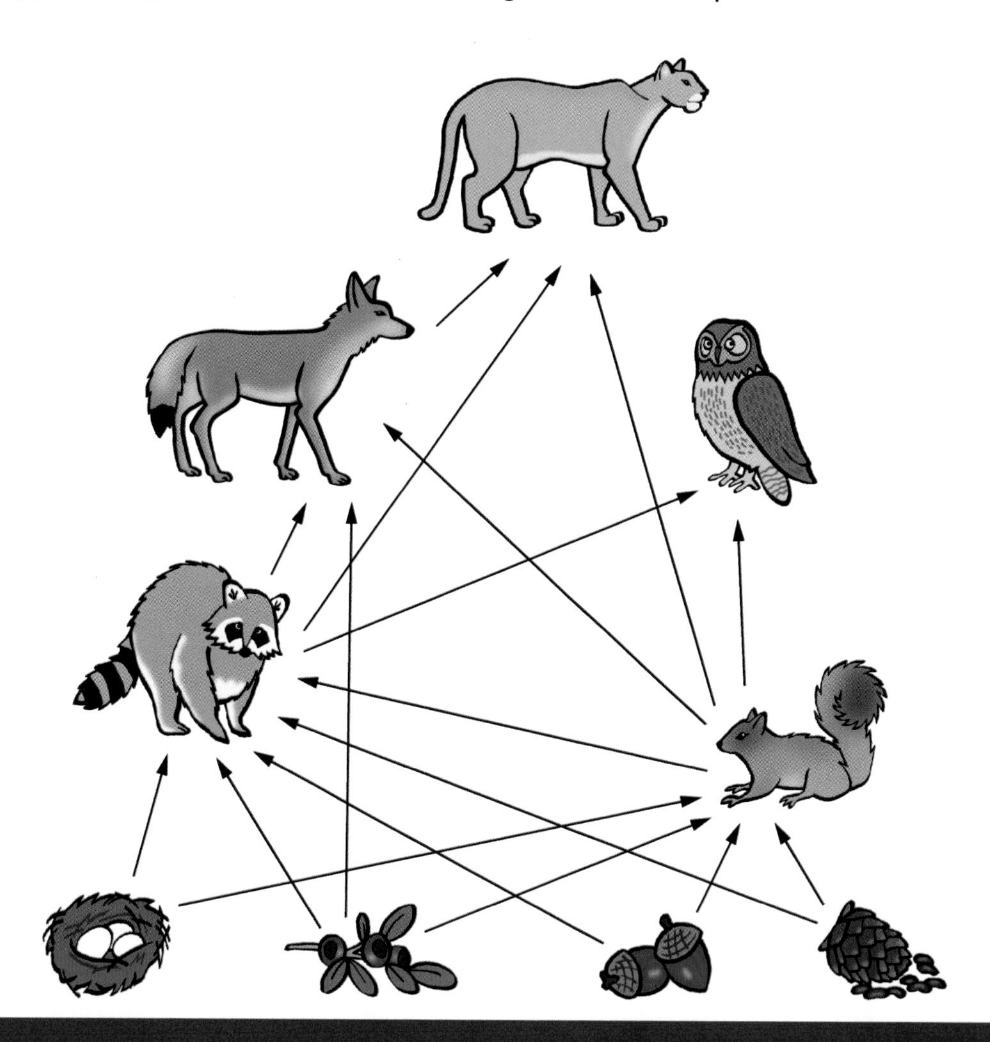

A food web

In this example, both squirrels and raccoons feed on eggs, berries, acorns, and pine nuts. However, raccoons will also eat squirrels. Coyotes and owls will both eat raccoons and squirrels. The mountain lion eats coyotes, raccoons, and squirrels and is an apex predator—there are no other animals that eat it.

10.5 Summary

- The biosphere is the part of Earth that is inhabited by living things. The biosphere goes up into the atmosphere and includes the land masses and the oceans from top to bottom.
- Some cycles that are important to the biosphere are the oxygen cycle, the carbon cycle, the nitrogen cycle, and the energy cycle.
- Ecology is the study of the relationship of living things to the world around them.
- An ecosystem is a specific area that contains a community of living things that exist under similar conditions.
- The place in an ecosystem where a particular organism lives is called its habitat.
- The role a species plays within an ecosystem is called its niche.
- A biome is a large ecosystem defined by the climate, soils, and plant life existing within it.
- A food web describes the interrelationships of how species feed on and are fed by each other, and tracks how energy moves through the organisms in an ecosystem.

10.6 Some Things to Think About

- In the area where you live, what are some changes to the biosphere that humans have made? How were living things affected?
- Explain one of the cycles presented in this section.
- In what ways do you think the oxygen cycle, the carbon cycle, the nitrogen cycle, and the energy cycle work together to support life?
- Take your field notebook and go for a walk. Find a living thing you think is interesting and observe the organism and its surroundings. Why is the organism living in that area? What necessary conditions are provided by the ecosystem? How would you describe its habitat and its ecosystem? What biome do you think it lives in? Record your observations.
- How do you think mapping food chains and food webs helps biologists study an ecosystem?
- What do food webs tell biologists that food chains don't?

Chapter 11 The Magnetosphere

Illustration of magnetosphere and Cluster satellites courtesy of ESA, AOES Medialab

11.1 Introduction

Scientists theorize that the movement of metals in Earth's core creates a magnetic field that surrounds Earth. This magnetic field can be detected with a compass that has a magnetized needle. But does the magnetic field extend out into space? If so, does it affect the area of space it exists in? Does space affect the magnetic field? Does the Sun have any interaction with the magnetic field? In this chapter, we will explore Earth's magnetic field, and we'll learn about the magnetosphere surrounding Earth.

11.2 Earth's Magnetic Field

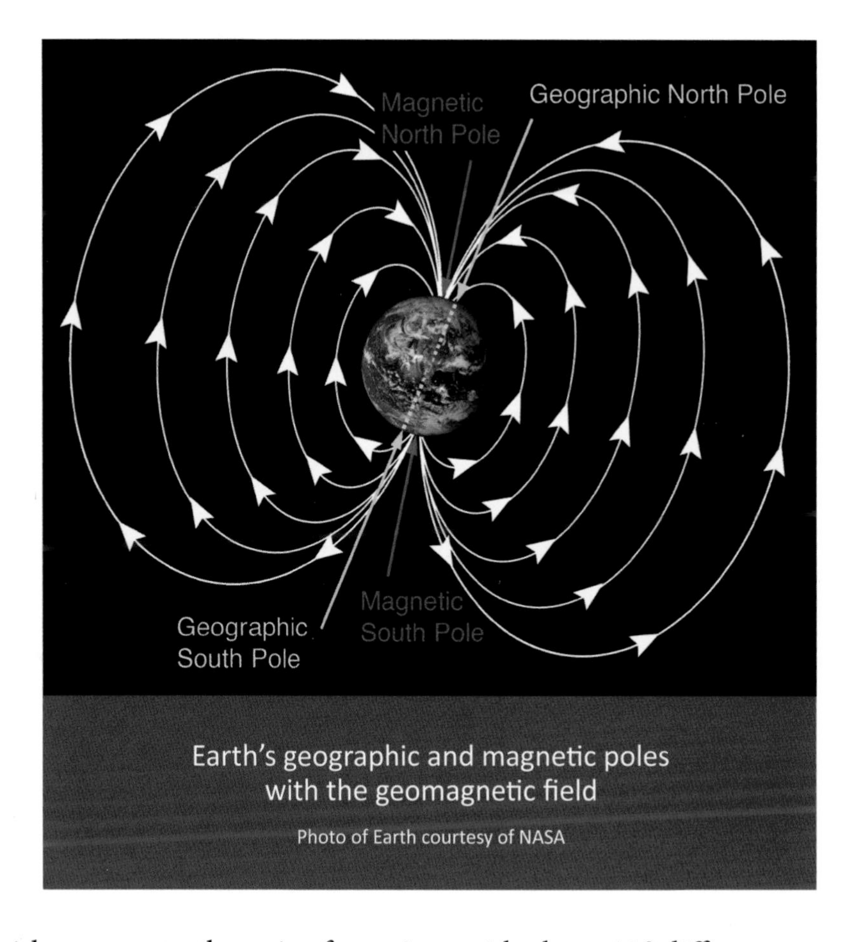

Earth's geographic and magnetic poles with the geomagnetic field

Photo of Earth courtesy of NASA

Earth is like a big magnet, and we can visualize this by imagining a huge bar magnet going through the center of the Earth. Just as a bar magnet has two poles, Earth has two magnetic poles, the Magnetic North Pole and the Magnetic South Pole, which are different from the geographic poles. The Geographic North Pole (also called true north) is the northernmost point of Earth's axis of rotation. The Magnetic North Pole is the place a compass points to and is one end of Earth's magnetic field. These two poles are not in quite the same location. Earth's magnetic field is tilted with respect to the axis of rotation, with about 11° difference between the location of the Geographic North Pole and the Magnetic North Pole. For this reason, compasses do not point directly toward the Geographic North Pole but rather to the magnetic pole. In the Southern Hemisphere, the Geographic South Pole and the Magnetic South Pole are also in different locations.

Earth's magnetic field, also called the geomagnetic field, is the region in which magnetic forces can be observed. The geomagnetic field is three-dimensional, totally surrounding the Earth. This is why a compass will work anywhere on Earth's surface.

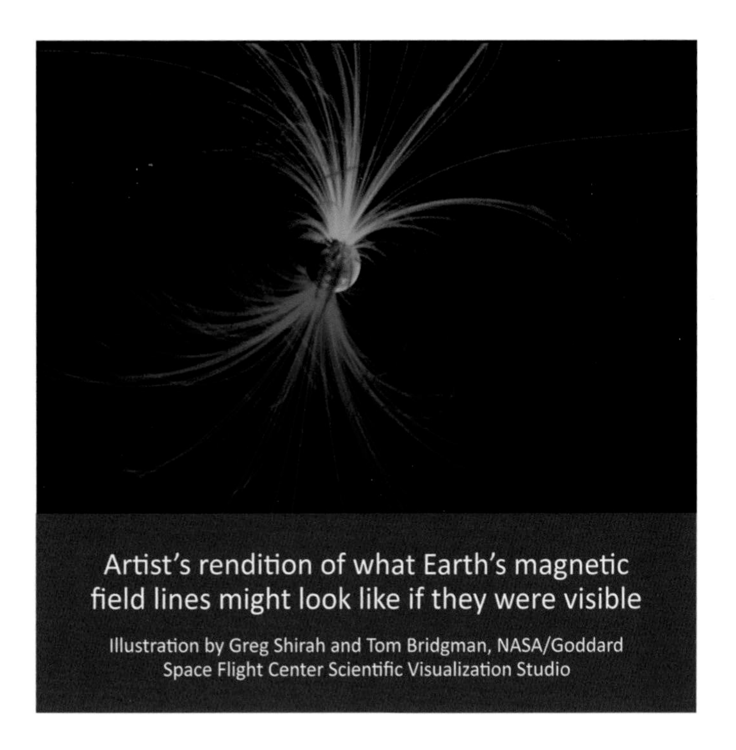
Artist's rendition of what Earth's magnetic field lines might look like if they were visible

Illustration by Greg Shirah and Tom Bridgman, NASA/Goddard Space Flight Center Scientific Visualization Studio

In illustrations of Earth's magnetic field, magnetic field lines are drawn to represent the direction of the magnetic force, and they show how a magnetized object will be affected at any point in the magnetic field. If we were to place a compass at a particular point on a magnetic field line, the field line would indicate in which direction the compass needle would be pointing if it were free to move in any direction. We can see in the illustration at the beginning of this section that if we follow the direction of a magnetic field line in the Northern Hemisphere, the compass will always be pointing north. Near the equator, the needle will be horizontal, or parallel to Earth's surface, as reflected by the magnetic field line. The closer we get to the Magnetic North Pole, the steeper the angle of the needle will become, until finally at the Magnetic North Pole, it will point straight up and down. In the Southern Hemisphere, it will be the south end of the compass needle that points downward as we move closer to the Magnetic South Pole. The north end of the needle will still be pointing toward the north.

11.3 Earth's Dynamo

Scientists think the geomagnetic field begins deep within the Earth, in the core. It is thought that the inner core is solid and has iron as its main component. The outer core is liquid and made mostly of the metals iron and nickel.

Scientists don't know exactly how Earth generates its magnetic field, but one theory is that the molten iron acts like a huge dynamo. A dynamo is a machine that converts mechanical energy into electrical energy. In physics we discover that electrical currents can be produced by moving a bar magnet through a coiled wire. In this way dynamos produce electrical current by rotating coils of wire through magnetic fields.

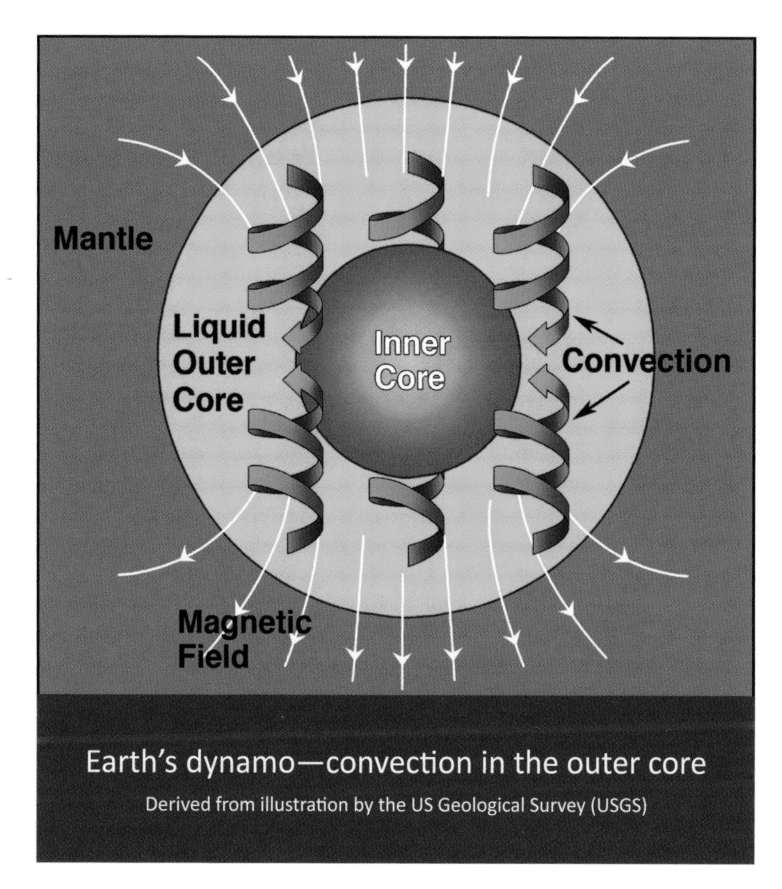

Earth's dynamo—convection in the outer core
Derived from illustration by the US Geological Survey (USGS)

To explain the creation of Earth's magnetic fields, scientists think that as Earth rotates, the molten iron in the outer core also rotates and, together with convection currents, creates a conducting fluid. This conducting fluid creates moving electrical currents around the rotational axis. When electrical currents travel in a loop, a magnetic field is induced. Scientists think that in this way the conducting fluid surrounding the inner core produces Earth's magnetic field.

11.4 Magnetic Field Reversals

Scientists have discovered that in the distant past Earth's geomagnetic field has reversed itself a number of times, with north becoming south and south becoming north! This change in the geomagnetic field is called a geomagnetic reversal. A geomagnetic reversal is thought to happen over a very long period of time with the magnetic field shifting position gradually. This shifting of the position of the geomagnetic field is thought to be due to changes in the flow of the molten metals in the Earth's outer core. During the 20th century the Magnetic North Pole moved about 1100 kilometers (684 miles). The strength of Earth's geomagnetic field also changes over time, getting stronger and weaker. Since about 1850 it has been weakening slightly.

Evidence of the effect of Earth's geomagnetic field is stored in ancient, metal-containing rocks. This allows scientists to determine what the position of the Magnetic North Pole was when the rocks were formed. When rock is in the molten stage, the magnetic fields of metallic elements such as iron and nickel line up with respect to the magnetic poles of Earth. As the rock cools, these metallic elements keep their magnetically induced alignment. By drilling out core samples of ancient rocks, scientists can use the magnetic field alignment of the metals in the different layers to tell where the Magnetic North Pole was located at different times in Earth's history.

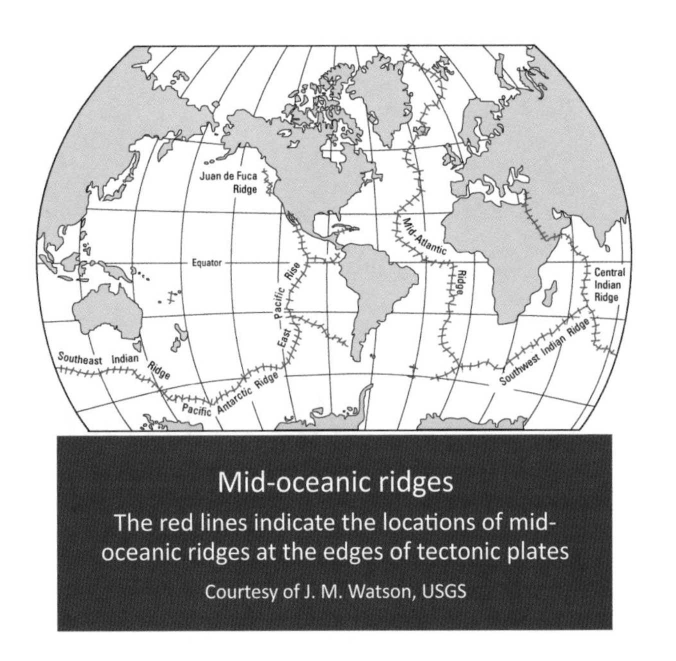

Mid-oceanic ridges

The red lines indicate the locations of mid-oceanic ridges at the edges of tectonic plates

Courtesy of J. M. Watson, USGS

In addition to core samples, scientists use rocks in the sea floor to map how Earth's magnetic field has changed over time. Recall from Chapter 5 that Earth's crust is broken up into tectonic plates that move slowly due to convection currents in the underlying molten asthenosphere. Some of the huge plates that are next to each other in the middle of the oceans are moving in opposite directions in a process called seafloor spreading. As the plates move apart, magma wells up in the crack between the two plates and forms a mountain chain called a mid-oceanic ridge. As the magma cools, new rock is created on both sides of the ridge at the same time. This is a continuous process with the plates moving apart and magma welling up in the crack and hardening into new rock.

In the 1950s scientists began mapping magnetic fields in the ocean floor by towing behind ships a magnetometer, an instrument that measures magnetic force. They were surprised to find that the magnetic fields in the rocks on the ocean floor are arranged in stripes that are parallel to the mid-oceanic ridge. Each stripe has the opposite polarity to the ones on either side of it. If you could put a compass on each stripe of rock, the needle would point toward the current Magnetic North Pole in one, and it would point toward the current Magnetic South Pole in the next. Another surprise was that the stripes on either side of the ridge had the same width and the same alternation of magnetic orientation between north and south. This led to the understanding that the stripes of rock on both sides of the oceanic ridge were formed at the same time. Recordings of the changing magnetic orientations in the rock in the ocean floor were compared to core samples from thick, old lava flows, and it was found that the magnetic field reversals corresponded in age and orientation in the two types of samples.

The variation in width of the stripes helps scientists determine how much time passed between magnetic field reversals. It has been found that geomagnetic pole reversals are unpredictable and appear to occur from anywhere between hundreds of thousands of years to a million years apart with the last pole reversal occurring about 780,000 years ago. To further their study, scientists have done computerized experiments simulating how they think Earth's dynamo system creates the magnetic field. In these simulations, pole reversals were randomly occurring events that were both natural and unpredictable.

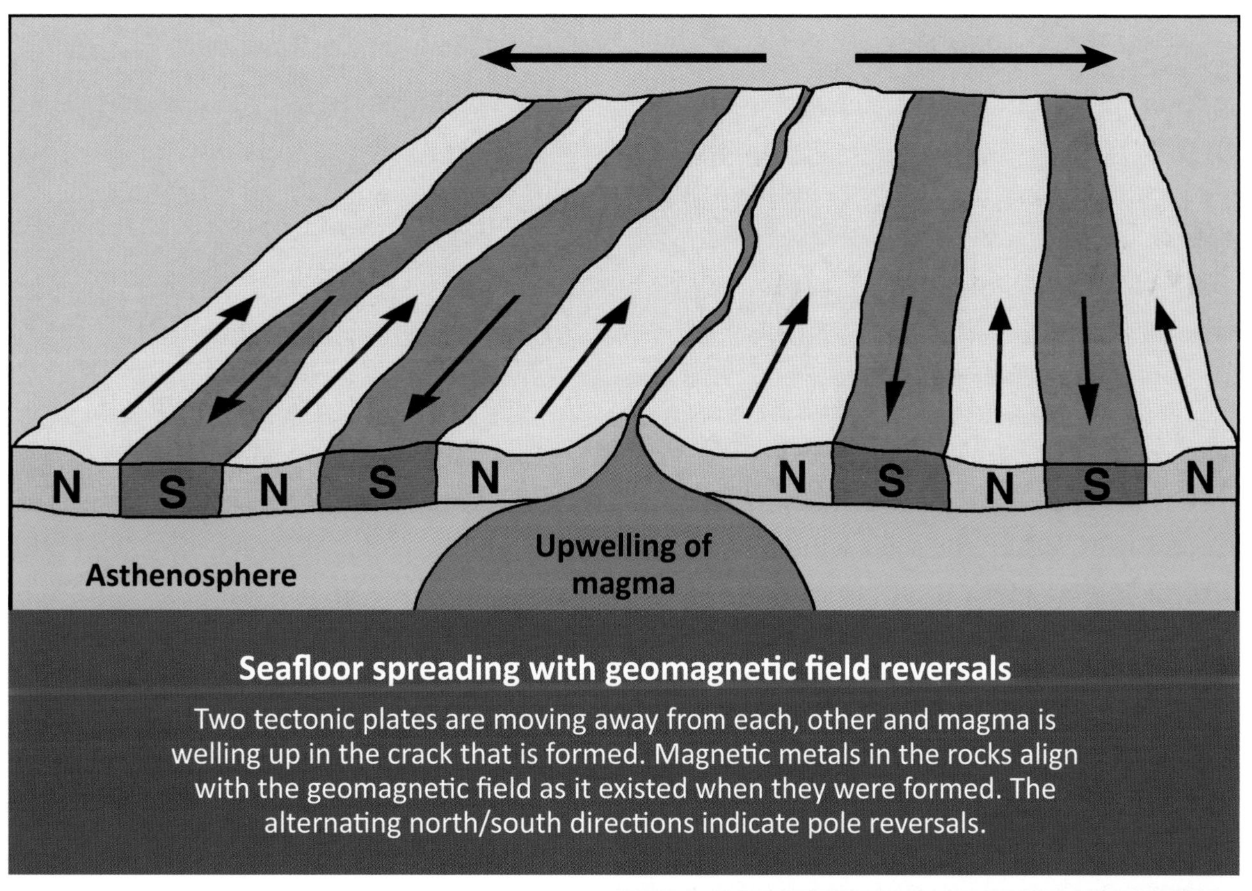

Seafloor spreading with geomagnetic field reversals

Two tectonic plates are moving away from each, other and magma is welling up in the crack that is formed. Magnetic metals in the rocks align with the geomagnetic field as it existed when they were formed. The alternating north/south directions indicate pole reversals.

The study of ancient rocks to learn about Earth's geomagnetic field is called paleomagnetism, and scientists who do these studies are called paleomagnetists. "Paleo" comes from the Greek word *palaios* which means ancient. Paleontologists study the fossil records (remains of ancient plants and animals) contained in the rocks. Using the

Fossils

Courtesy of Mark A. Wilson, The College of Wooster

information gained through paleomagnetism and paleontology, scientists have concluded that geomagnetic reversals do not significantly affect life on Earth. There's no evidence that massive die-outs of life have occurred as a result of geomagnetic pole reversals.

11.5 Earth's Magnetosphere

Surrounding Earth is the magnetosphere, the area in space that contains Earth's magnetic field. The magnetosphere is formed by the interaction of the geomagnetic field with gases coming from the Sun. Most of this gas is deflected by the magnetosphere, protecting Earth from harmful radiation. The magnetosphere is bullet shaped with the curved end pointing toward the Sun, and it is made up of different layers that have different functions. With the use of satellites, scientists are beginning to discover how the different layers work together and what they do, but much is still unknown.

The Sun constantly emits a massive stream of rapidly moving hot gas called plasma. Plasma is made up of electrically charged particles (negatively charged electrons and positively charged protons and ions) and is considered by some scientists to be the fourth state of matter. Solar wind, the massive stream of plasma coming from the Sun, continuously speeds toward Earth at about 400 km/second (about 1 million miles per hour), although the speed varies. When the solar wind nears Earth's geomagnetic field, it is largely blocked and diverted by magnetic forces from Earth. This diversion causes the solar wind to flow around the geomagnetic field, forming the bullet shape of the magnetosphere. A way to visualize this is to think of a rock in a stream of water and imagine how the water is diverted by the rock and flows around it. Or imagine a boat moving through water with the bow of the boat diverting the water that then moves around the boat.

Closest to the Sun is the arc shaped bow shock, the area in space where the solar wind is suddenly slowed as it encounters the forces of the geomagnetic field. Inside the bow shock, closer to Earth, is the magnetosheath which is the transition area between the bow shock and the magnetosphere itself. Solar winds in the magnetosheath that have been slowed by the bow shock are deflected by the magnetosphere and slide around Earth's geomagnetic field, pulling the magnetosphere out into a long tail behind Earth. This long tail is called the magnetotail and is found extending out from the side of Earth that is away from the Sun (the night side). The magnetotail is so long that its length hasn't been measured, but an estimate is several hundred thousand kilometers.

Within the magnetosheath is the magnetopause, the outer boundary of the magnetosphere. At the magnetopause the pressure from the solar wind is equal to the pressure of the geomagnetic field. This keeps most, but not all, of the solar wind plasma from entering the magnetosphere. The pressure of the solar wind pushing on the part of the magnetopause that is facing the Sun is strong enough to squash the geomagnetic field in this area, giving it a somewhat flattened shape on the Sun, or day, side. Although the magnetotail on the night side of Earth is extremely long, the distance from Earth to the bow shock on the day side is only about 60,000 km (40,000 miles).

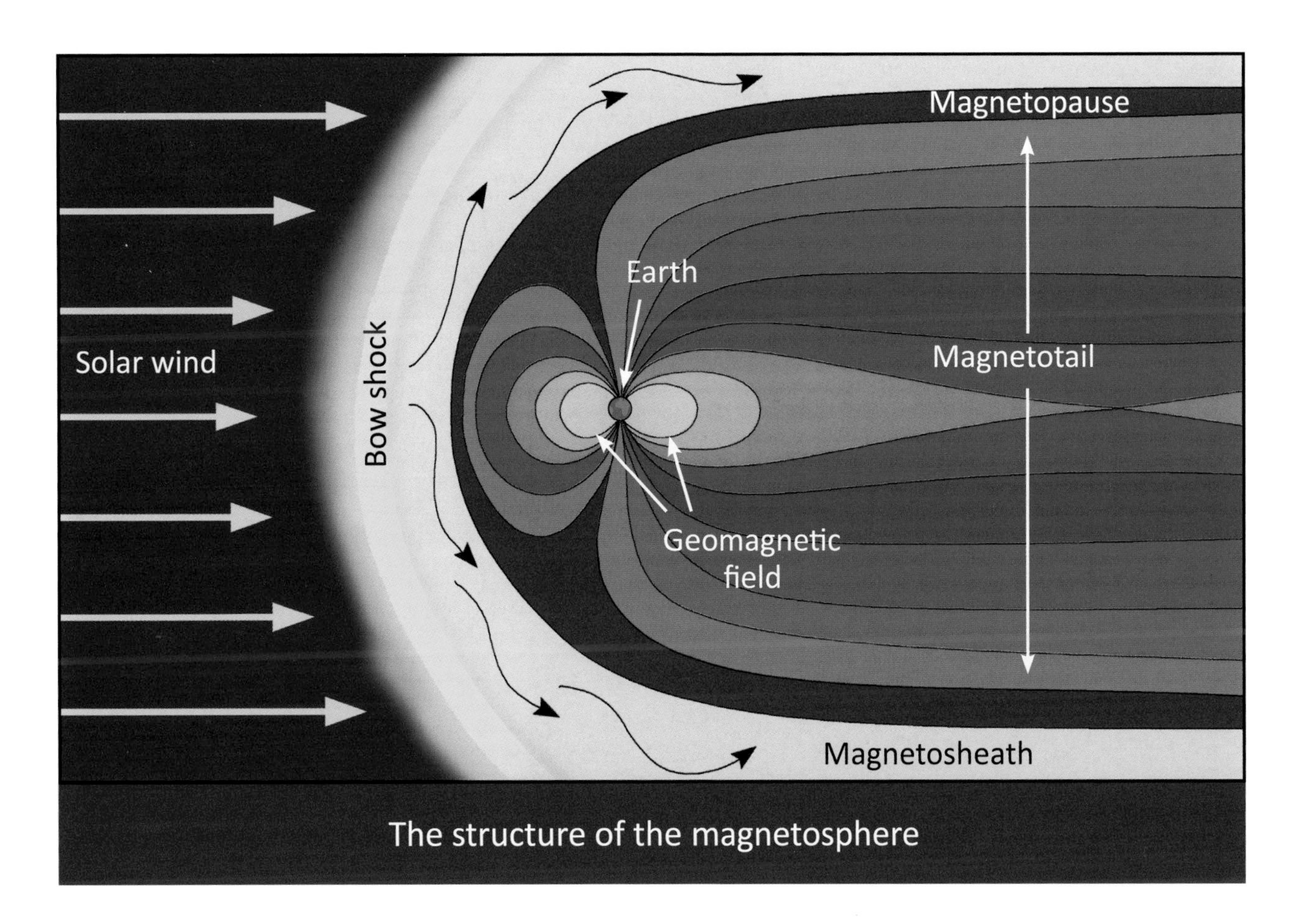

The structure of the magnetosphere

The magnetotail described earlier is found within the magnetopause and is made up of several layers. Within the magnetotail, the geomagnetic field on the night side of Earth is stretched out into a teardrop shape. Some of the plasma from the solar wind makes its way into the magnetotail where the particles are stored. Current theory is that electrons stored in the magnetotail can become energized and will then travel along magnetic field lines from the magnetotail to the polar regions of Earth. Here the electrons collide with atoms and molecules of oxygen and nitrogen in the atmosphere. These collisions cause the electrons to release energy in the form of light, resulting in a beautiful display of moving curtains or streamers of color called an aurora. Collisions with oxygen molecules cause green colors to be produced, and nitrogen collisions result in blues and reds. An aurora that occurs near the Magnetic North Pole is called the northern lights, or aurora borealis, and one near the Magnetic South Pole is called the southern lights, or aurora australis. *Aurora* is the Latin word for dawn.

An aurora borealis in Alaska
Courtesy of NASA/US Air Force, photo by Senior Airman Joshua Strang

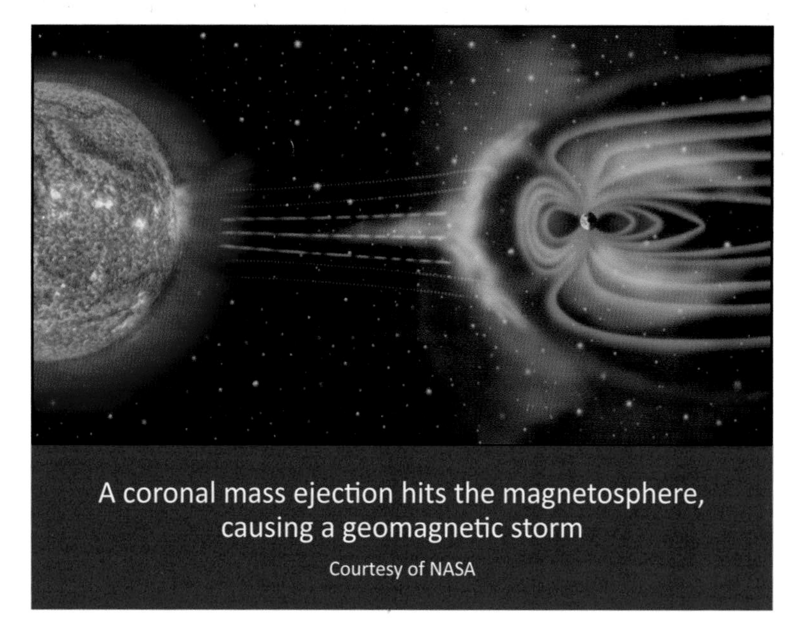

A coronal mass ejection hits the magnetosphere, causing a geomagnetic storm
Courtesy of NASA

The Sun also has a magnetic field and from time to time has solar storms. One type of solar storm happens when changes in the Sun's magnetic field cause a coronal mass ejection, or CME. During a coronal mass ejection, a huge cloud of plasma is blasted from an area of the Sun. If the CME occurs in a region of the Sun that is pointed toward Earth, the cloud of fast moving plasma can hit the magnetosphere, causing a geomagnetic storm. The disturbance of the magnetosphere during a geomagnetic storm can cause more charged plasma particles than usual to enter the magnetotail, resulting in auroras that are much larger than usual and that can be seen from locations farther from the poles. It is thought that, in addition to more plasma entering the magnetotail, a CME can cause charged particles from Earth's upper atmosphere to be released into space.

Also, the force from a CME compresses the front of the magnetosphere more than usual. This can cause problems for satellites if they find themselves outside the protective magnetosphere, and they can be damaged by harsh solar radiation. And a geomagnetic storm can cause interruptions in radio and radar transmissions, as well as problems with the power grids on Earth, resulting in disruptions in the delivery of electrical power.

The magnetosphere is a dynamic area in space. Its shape is constantly adjusting as it interacts with changing pressures created by the varying strength, speed, and density of the solar winds that cause plasma to enter and be released by the magnetosphere. As Earth rotates around the Sun, the magnetosphere always stays in the same orientation, with the bow shock toward the Sun and the magnetotail pointed away from the Sun. Because the magnetosphere keeps Earth from receiving excessive amounts of energy from the Sun, life could not exist on Earth without it.

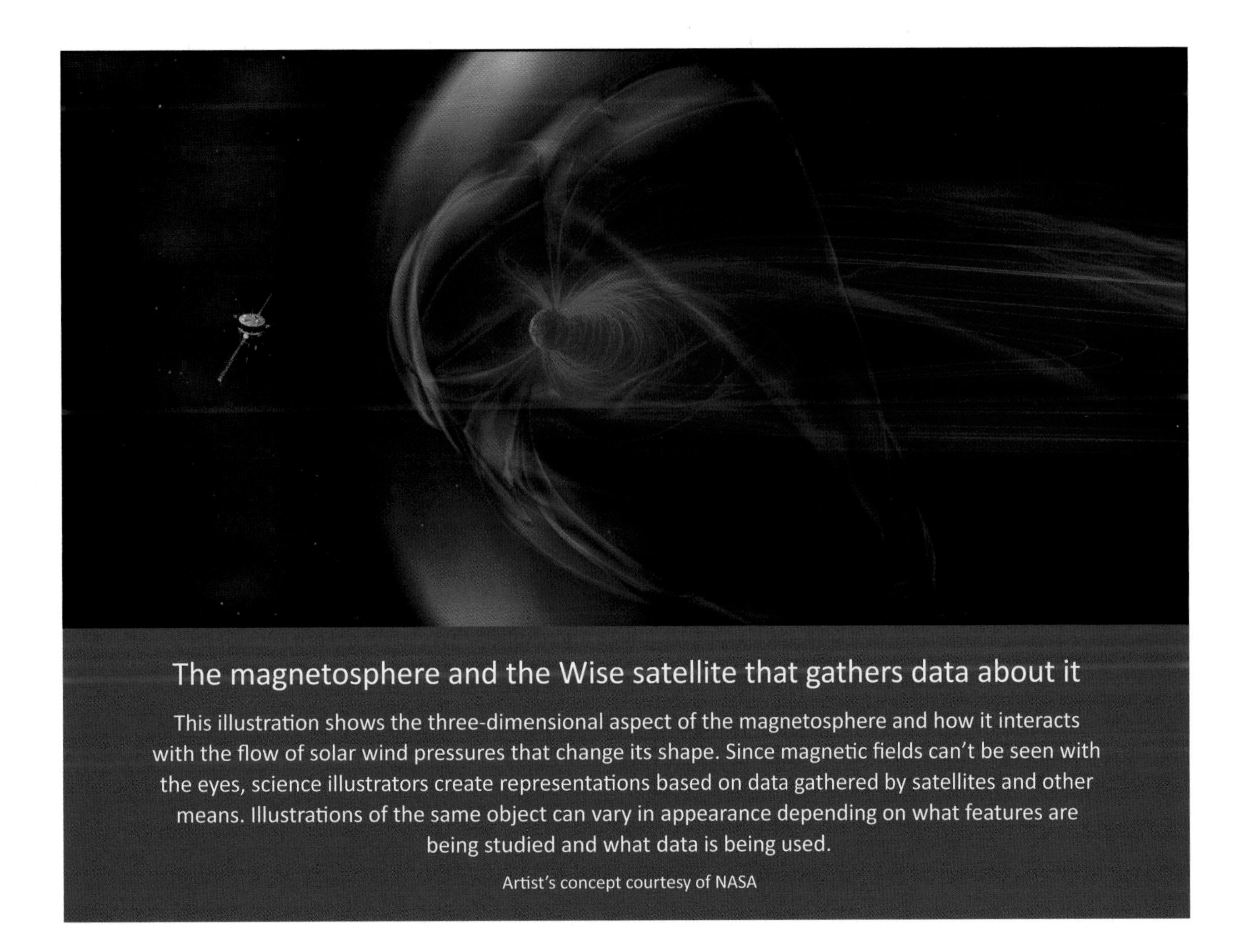

The magnetosphere and the Wise satellite that gathers data about it

This illustration shows the three-dimensional aspect of the magnetosphere and how it interacts with the flow of solar wind pressures that change its shape. Since magnetic fields can't be seen with the eyes, science illustrators create representations based on data gathered by satellites and other means. Illustrations of the same object can vary in appearance depending on what features are being studied and what data is being used.

Artist's concept courtesy of NASA

11.6 Summary

- Earth has two magnetic poles, the Magnetic North Pole and the Magnetic South Pole, which are at slightly different locations than the Geographic North and South Poles.
- Earth's geomagnetic field is thought to be created by electrical forces that are formed by the movement of molten metals in the outer core.
- The magnetosphere is created by the interaction of Earth's geomagnetic field with gases from the Sun.
- Periodically, the Magnetic North Pole and the Magnetic South Pole undergo a geomagnetic reversal, with north becoming south and south becoming north.
- The magnetosphere is the area in space that contains Earth's magnetic field and is formed by the interaction of the geomagnetic field with gases coming from the Sun.
- Solar wind is the steady stream of hot gases, or plasma, that comes from the Sun.
- Paleomagnetism is the study of ancient rocks to learn about Earth's geomagnetic field in the past.

11.7 Some Things to Think About

- The magnetized needle of a compass always points north. How do you think you could use a compass to find south, east, and west?
- In your own words, explain the theory of how Earth's magnetic field may be created. Do you think this theory is probable? Why or why not?
- What do you think the effects of a magnetic pole reversal might be? Why?
- What features of the magnetosphere do you find most interesting? Why?

Chapter 12 Earth as a System

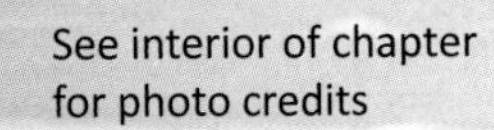
See interior of chapter for photo credits

12.1 Introduction

The Earth operates as a system. It has many parts that make up the whole. The biosphere, hydrosphere, atmosphere, magnetosphere, lithosphere, asthenosphere, and core can each be studied separately, but all of these spheres influence each other. For example, when it rains, the hydrosphere is interacting with the biosphere, helping plants and animals to grow. When a hurricane builds in the ocean, the atmosphere is interacting with the hydrosphere and both then interact with the biosphere and lithosphere when the hurricane reaches land. A volcano begins in the mantle, breaks through the lithosphere, and interacts with the atmosphere and biosphere as it erupts.

Interdependent spheres

Derived from illustrations by the US Geological Survey and Max-Karl Winkler, National Science Resources Center

12.2 Interdependence

The different spheres on Earth are interdependent which means they interact with and rely on each other in order to function properly. A tropical rainforest is an example of interdependence. The trees, bushes, and animals are part of the biosphere and exchange carbon dioxide and oxygen with the atmosphere. The atmosphere carries water from the hydrosphere, providing rainfall, which enters the soil in the lithosphere where it is taken up by the roots of the plants in the biosphere. The atmosphere also provides nitrogen to the soil in the lithosphere where it is fixed by bacteria which are part of the biosphere. The atmosphere and magnetosphere allow the proper amount of energy from the Sun to fall on the rainforest so the plants can use it for photosynthesis. Even in this simplified view of the rainforest ecosystem, it's easy to see how complicated and essential the interdependence of the various spheres is.

A house in the Amazon rainforest biome
Courtesy of Francisco Chaves from Buenos Aires, Argentina, CC BY SA 2.0

Imagine how a change in one of the spheres might affect the entire tropical rainforest. For example, the trees might die if the amount of rainfall from the atmosphere changed and the climate became too wet or too dry. If the atmosphere or the magnetosphere let in too much energy from the Sun, the trees might burn up; too little and they might not have enough energy to make sufficient food. If the trees died, birds and animals would not have food or places to live.

Saharan dust blows across the Atlantic to the Amazon
Courtesy of NASA, image created by Jesse Allen, Earth Observatory, using data obtained courtesy of the MODIS Rapid Response team

Individual ecosystems are also dependent on parts of Earth that are far away. The ocean currents that begin near the North Pole circulate all the way around the globe, affecting the land masses by creating cooling and warming winds and providing moisture that falls as precipitation. Winds blow dust from the African Sahara Desert all the way across the Atlantic Ocean where it falls on the Amazon rainforest, delivering much needed nutrients. Monarch butterflies migrate from Mexico and pollinate plants in locations as far away as 4800 km (3000 miles).

We can begin to see what an intricate task it is for Earth to keep all its systems in balance and how difficult it is to try to figure out all the ways the different spheres interconnect and affect each other.

12.3 Natural Events

A hurricane in the Gulf of Mexico
Courtesy of NASA/GOES Project

Humans have no control over natural events such as volcanic eruptions, earthquakes, hurricanes, and tsunamis. These events can affect many spheres of the Earth system and can have devastating effects on people. For example, a volcanic eruption can change the lithosphere with deposits of lava, mud flows, and thick layers of ash. These deposits can change the biosphere when they cover and kill plants, animals, and people. A big eruption can fling large quantities of ash into the atmosphere where it can block energy from the Sun, thus reducing the amount of energy available for photosynthesis in plants in the biosphere. Ash in the air can also affect the hydrosphere by changing the amount of rain that will fall later. Some scientists think that huge volcanic eruptions in the distant past led to the extinction of some species.

Although we can't control natural events, scientists continue to study things such as volcanoes, earthquakes, hurricanes, and tsunamis. The expectation is that someday we will be able to predict these events far enough in advance that people can be prepared.

12.4 Changing the Carbon Cycle Changes Earth

As we observe both our local climate and the global climate from year to year, it seems clear that the climate is changing. In many places temperatures are warmer in both summer and winter. Changes in climate make it more unpredictable when and how much precipitation will arrive, and storms are bigger, wilder, and more destructive. Species are becoming endangered as their habitats become too warm or too dry, and all over the world glaciers are melting.

In this section we will offer brief summaries of a few of the many scientific findings about climate change and the data that supports current theories. Although the majority of scientists agree that human activities are, in fact, coupled to climate change, there is still some debate about whether human activity affects the global climate and if so, how much effect there is and what to do about it. Because Earth's spheres are so huge and their interactions are extremely complicated, it can be difficult for scientists to come up with definitive answers, and theories change as new discoveries are made.

Regardless of what you might hear or believe about climate change, if you have some understanding of the scientific research that has been done, you will be better able to form your own opinions when you hear discussions about climate change. There is an amazing amount of scientific work being carried out to study this subject. To find out facts in addition to what is presented here, you can research reputable science-based organizations that are doing research projects. Some United States government agencies that have websites where you can find more information are the US Geological Survey (USGS), the National Oceanic and Atmospheric Administration (NOAA), the National Aeronautics and Space Administration (NASA), and the Environmental Protection Agency (EPA), among others. You can also do an internet search for agencies in other countries and for universities that study climate change. As with anything on the internet, it is very important to look for reliable, credible sources of information—people who are actually doing research and have data rather than people who are just stating opinions without a factual basis for their ideas.

Global Climate Change

The term climate refers to the typical or average weather in a certain region over a long period of time, including factors such as precipitation and temperature. For example, you may live in an area that on average is warm and sunny or one that is cool and cloudy. The region can be defined as anywhere from a city to the Earth as a whole. When doing research to find out whether the climate is changing in a particular region, scientists collect data about factors such as changing weather patterns and average temperatures and look at how these factors are affecting the land and living things. When considering the climate of Earth as a whole, researchers are looking at average global changes in weather patterns, average global temperature, and so forth.

One aspect of global climate change that is being studied by scientists is the trend toward warmer temperatures in many regions. Using a variety of instruments, scientists have been collecting data about Earth's spheres for many decades. One measurement being made is the worldwide average temperature of Earth. It has been discovered that the yearly average global temperature of Earth is increasing—Earth is getting warmer. Scientists are trying to figure out whether the increase in global temperatures is man-made or natural. Many suspect that it is man-made and that the major cause is the burning of fossil fuels. Beginning with the Industrial Revolution in the 1700s, the use of machinery driven by fossil fuels has increased, and global CO_2 emissions have increased as well, going from 280 parts per million* in the 1700s to about 400 parts per million in 2015, which represents an increase in CO_2 of over 40%! This increase in CO_2 emissions could be responsible for Earth's temperatures rising due to the increasing levels of CO_2 in the atmosphere adding to the greenhouse effect.

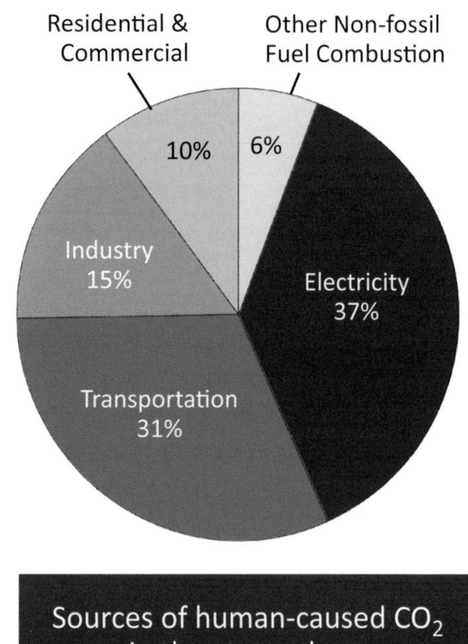

Sources of human-caused CO_2 in the atmosphere

Data as of 2014 from US Environmental Protection Agency (EPA)

The theory of the greenhouse effect was first proposed in 1824 by Jean-Baptiste Joseph Fourier (1768-1830), a French mathematician and physicist. Recall that greenhouse gases such as carbon dioxide absorb heat from the Sun and have a warming effect on Earth. Once the effect of greenhouse gases on global temperatures was known, scientists began to wonder if the ever-increasing amounts of CO_2 in the atmosphere due to the burning of fossil fuels could contribute to climate change by affecting global temperatures.

How does this work? Recall from Chapter 10 that carbon can enter the atmosphere quickly through the fast carbon cycle or more slowly through the slow carbon cycle. The fast carbon cycle includes the biosphere, hydrosphere, and atmosphere, and the slow carbon cycle includes rocks and organic matter found in the lithosphere. Both the fast carbon cycle and the slow carbon cycle can be thought of as being carbon-containing reservoirs, or pools, that take in and release carbon as part of the carbon cycle. What scientists are discovering

*Parts per million (ppm) refers to the concentration of a substance in a solution. The number 280 ppm of CO_2 means there are 280 kg of CO_2 in 1 million kg of air; 400 ppm is 400 kg of CO_2 in 1 million kg of air. Any unit of measurement can be used but the ratio remains 1 to 1 million.

is that CO_2 from fossil fuels is currently being released faster than it is being taken up by plants, land, and the ocean. Because CO_2 is a greenhouse gas, it absorbs heat from the Sun, warming the Earth, and as more CO_2 is released into the atmosphere, the average global temperature of Earth appears to be warming. If we look at carbon dioxide levels and global temperature we can see that both are trending upward.

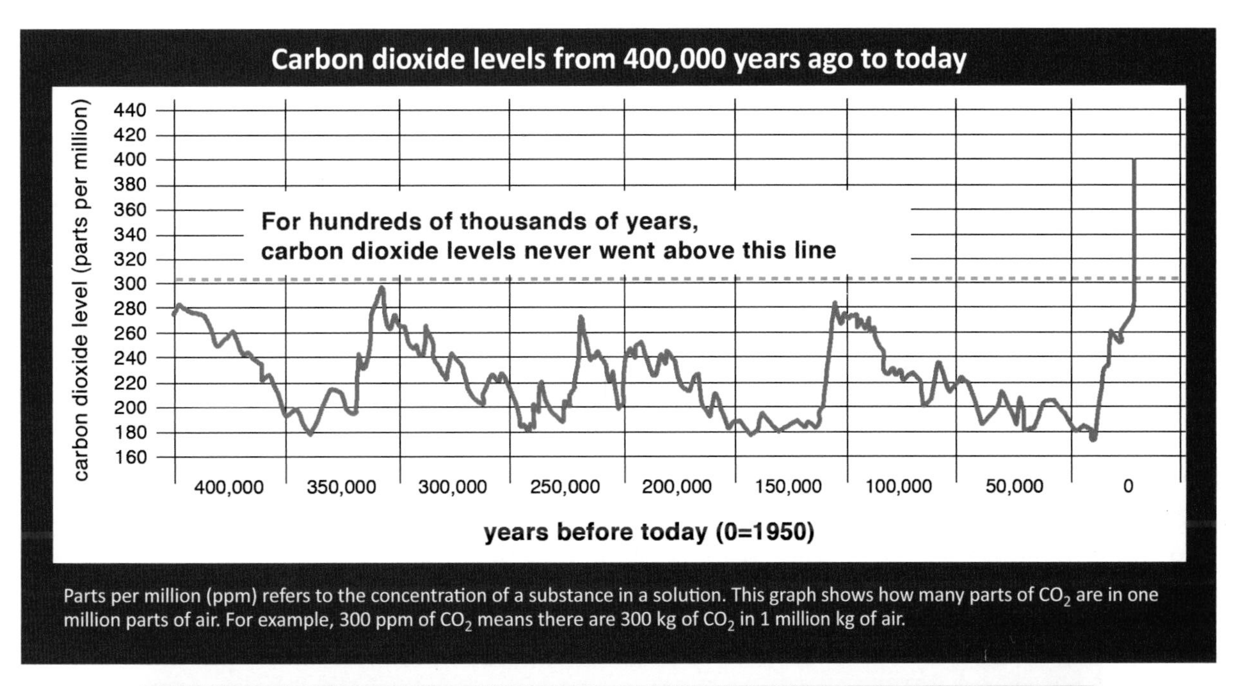

Parts per million (ppm) refers to the concentration of a substance in a solution. This graph shows how many parts of CO_2 are in one million parts of air. For example, 300 ppm of CO_2 means there are 300 kg of CO_2 in 1 million kg of air.

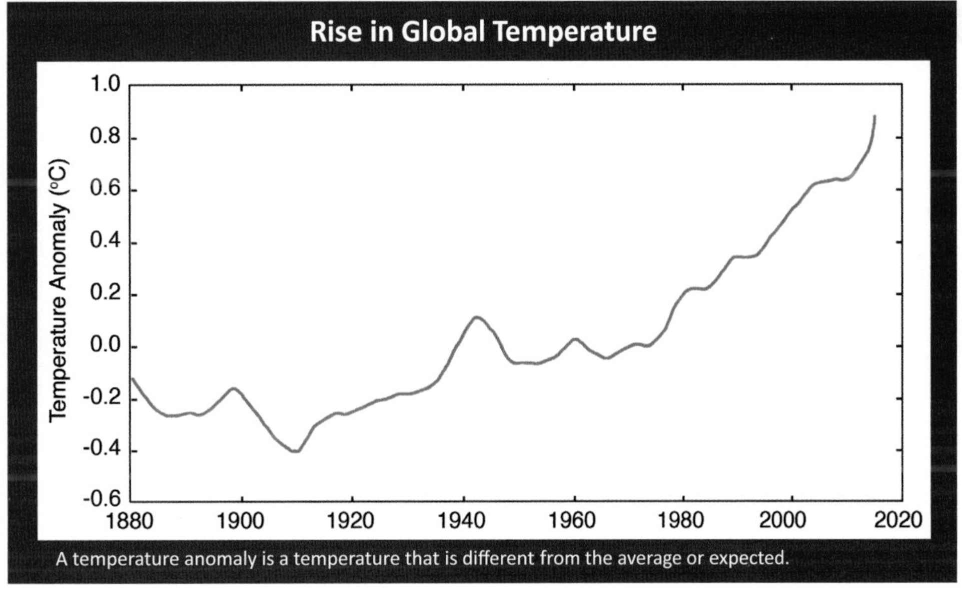

A temperature anomaly is a temperature that is different from the average or expected.

Changes in global temperature are also reflected in changing weather patterns. Heat waves in the summer are getting longer and hotter, leading to drought in some areas, stressing plants and animals and contributing to wildfires. Spring is coming an average of 10 days earlier in the Northern Hemisphere, affecting animal and bird migrations. Snow packs are melting earlier, which can cause less water to be available to plants and animals later in the summer, stressing plants and leaving animals thirsty. Sea levels are rising as the oceans become warmer, causing the seawater to expand and increase in volume, and polar ice is melting, adding more water to the oceans. The trees in some forest areas are dying as they become stressed by warmer, drier conditions and are more susceptible to insect damage. Some corals are dying as the oceans become warmer. We are seeing more extreme storms, such as hurricanes, that do more damage. These are a few of the changes we are already seeing that are expected to increase more rapidly as the average global temperature warms.

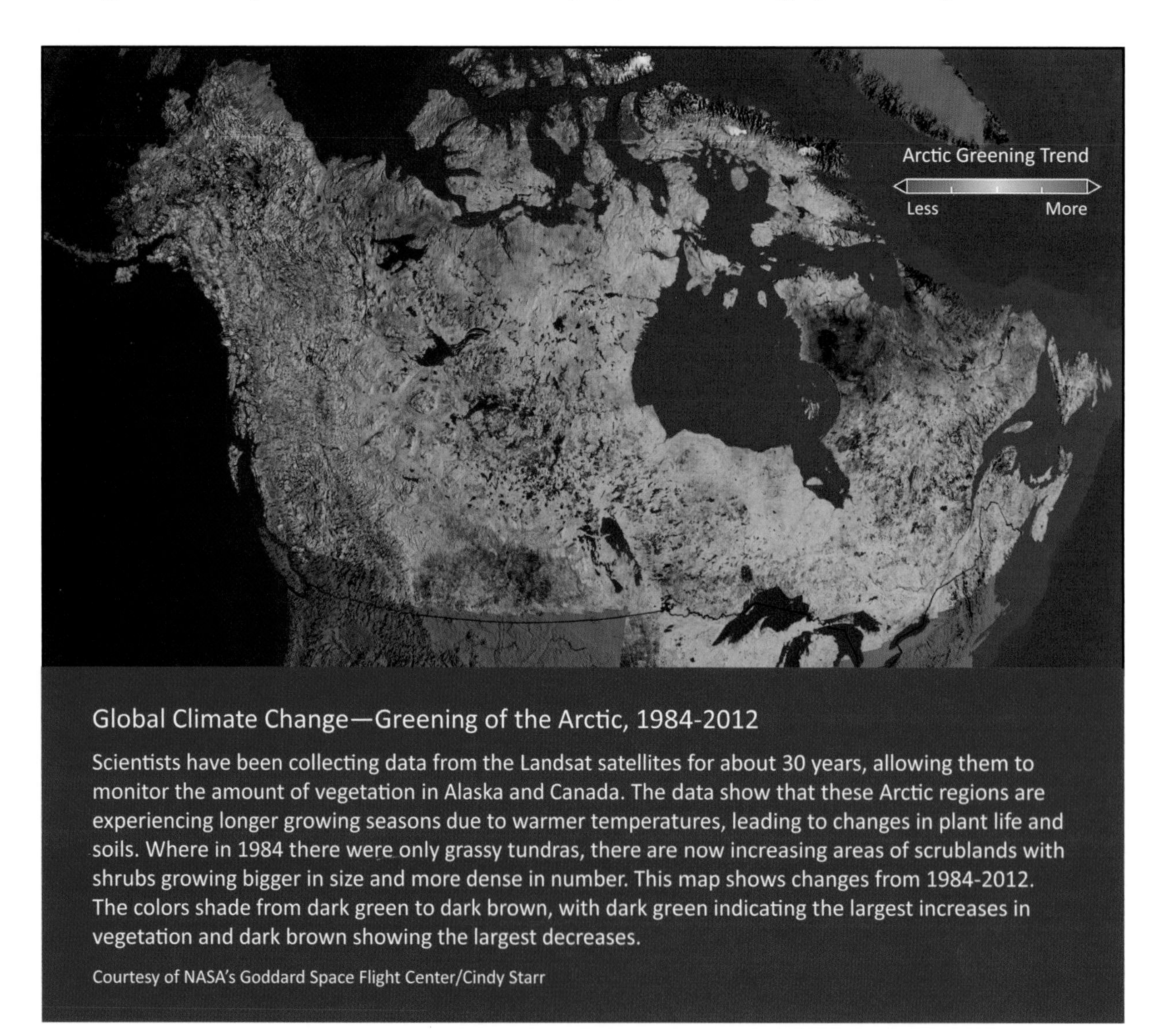

Global Climate Change—Greening of the Arctic, 1984-2012

Scientists have been collecting data from the Landsat satellites for about 30 years, allowing them to monitor the amount of vegetation in Alaska and Canada. The data show that these Arctic regions are experiencing longer growing seasons due to warmer temperatures, leading to changes in plant life and soils. Where in 1984 there were only grassy tundras, there are now increasing areas of scrublands with shrubs growing bigger in size and more dense in number. This map shows changes from 1984-2012. The colors shade from dark green to dark brown, with dark green indicating the largest increases in vegetation and dark brown showing the largest decreases.

Courtesy of NASA's Goddard Space Flight Center/Cindy Starr

Ocean Acidification

In addition to the Earth becoming warmer, the rising level of atmospheric CO_2 is leading to ocean acidification—the increasing acidity of seawater. As we saw in Chapter 9, the action of winds and waves mixes carbon dioxide from the atmosphere into the upper layer of the ocean, and ocean currents bring CO_2 from the upper layer into the deep ocean water. At one time scientists thought that the oceans would be able to absorb all the extra CO_2 being emitted by human activities, and the increasing CO_2 emissions would not be much of a problem. Also, the oceans are so huge that it was thought they would not be affected by more CO_2 being dissolved in the seawater. However, scientists now think that the oceans absorb only about 30% of human-caused carbon emissions and are becoming more acidic due to chemical reactions.

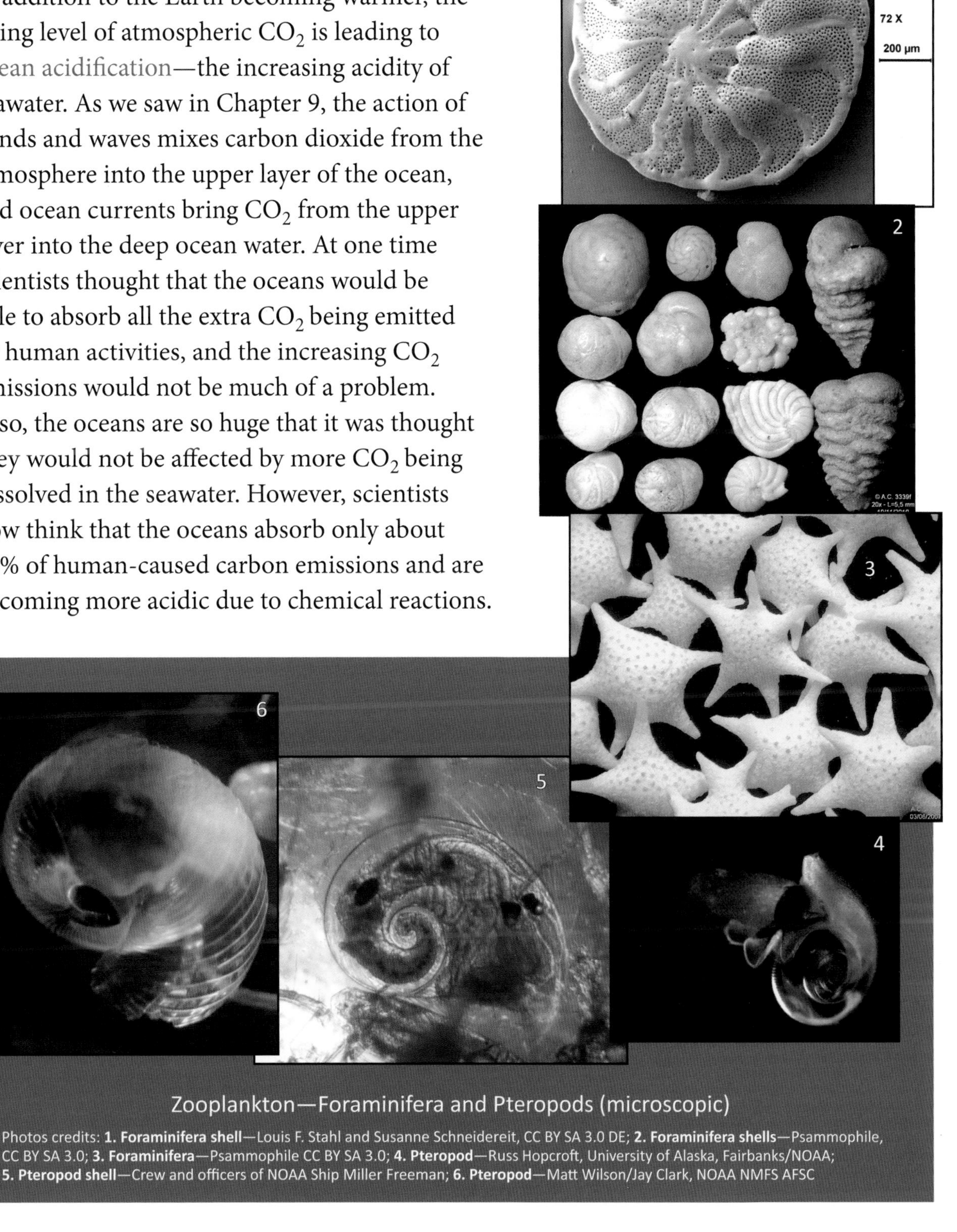

Zooplankton—Foraminifera and Pteropods (microscopic)

Photos credits: **1. Foraminifera shell**—Louis F. Stahl and Susanne Schneidereit, CC BY SA 3.0 DE; **2. Foraminifera shells**—Psammophile, CC BY SA 3.0; **3. Foraminifera**—Psammophile CC BY SA 3.0; **4. Pteropod**—Russ Hopcroft, University of Alaska, Fairbanks/NOAA; **5. Pteropod shell**—Crew and officers of NOAA Ship Miller Freeman; **6. Pteropod**—Matt Wilson/Jay Clark, NOAA NMFS AFSC

Zooplankton are tiny, floating marine animals that, along with phytoplankton, are a major food source for many animals and are at the base of the ocean food web. Foraminifera and pteropods are two types of zooplankton that build shells. It has been discovered in some areas of the ocean that due to ocean acidification the shells of these animals are dissolving or the animals are unable to build shells, causing the animals to die. Scientists are concerned that if too many of these tiny organisms die, the ocean food web will begin to collapse because there will be less food available to each successively higher trophic level.

An underwater sensor

An example of an underwater sensor used to detect CO_2 levels in water and water temperature. The data can be analyzed to study ocean acidification.

Courtesy of Bernadette Charpentier, NOAA

The full effects of ocean acidification on marine organisms are not yet well understood. Different organisms will react differently, with some more able to survive increased acidity than others. The ocean is vast and complicated, and there is still much to learn about its ecosystems and chemistry.

12.5 Endangered Species

Plant, fungi, and animal species whose numbers are declining to the point where they are in danger of becoming extinct are called endangered species. It is estimated that somewhere between 10-25 species become extinct each year through natural causes and that about the same number of new species develop to take their place. However, current estimates state that species are becoming extinct 1,000-10,000 times faster than this and are not being replaced by new species. Some scientists believe that we may be facing a mass extinction as great as the one that wiped out the dinosaurs! Some also predict that 30% to 50% of all species on Earth could become endangered by 2050 if humans don't begin to change our activities.

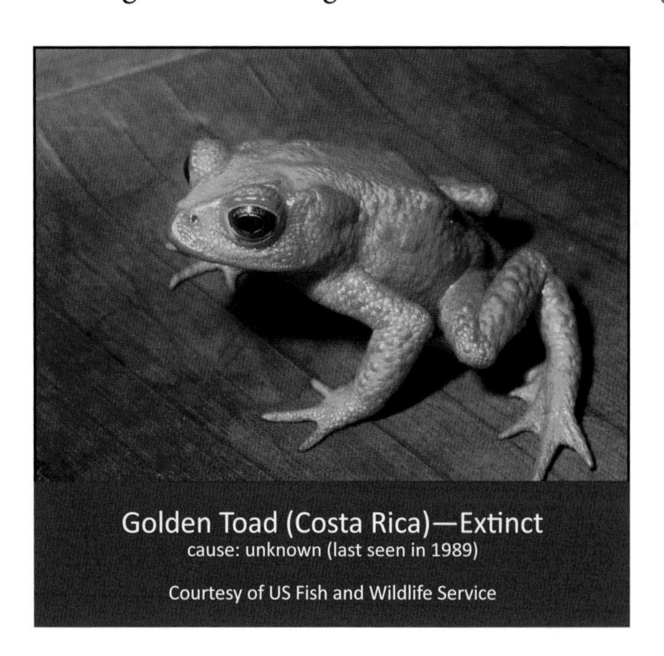

Golden Toad (Costa Rica)—Extinct
cause: unknown (last seen in 1989)

Courtesy of US Fish and Wildlife Service

Asian Elephant—Endangered
cause: illegal hunting and habitat loss

We know that global climate change is disrupting some ecosystems and habitats. Some species are able to adapt or move and some plants may thrive with more atmospheric CO_2 available, but those plant and animal species that cannot adapt quickly enough to changing habitats are in danger of becoming extinct. As human population growth occurs, ecosystems are impacted or destroyed by activities in addition to the burning of fossil fuels, such as land being cleared of vegetation, buildings constructed, roads built, and wetlands removed. Humans also pollute the land and water with chemicals and trash that can lead to drastically reduced or weakened animal populations. Also, in some areas humans have introduced invasive, nonnative species of plants and animals that sometimes crowd out native species or kill them through predation or introduced diseases.

It may not seem important if one species disappears, but we now know that just as all of Earth's spheres are interconnected, all living things and all ecosystems are interconnected. The removal of even one element of a food chain may not make much difference to an ecosystem, or it can drastically change or destroy it. If one ecosystem collapses, those nearby can also be affected. Our own survival as a species depends on the plants and animals that surround us, and we don't know at what point the destruction of individual ecosystems could lead to a larger collapse that would affect the entire Earth.

Pacific Coho Salmon—Endangered
human and natural causes

Courtesy of NOAA

We can view organisms that are facing extinction as early warning signals that something is going wrong in an environment. They indicate that something is becoming unhealthy in their ecosystem. By studying the factors that are impacting the health and survival of endangered species, we can learn more about what we can do to keep the planet healthy and to ensure continuation of our own species as well as that of the endangered plants, fungi, and animals.

Wercklea flavovirens (Jamaica)
Critically endangered
cause: human—habitat loss
Courtesy of Chiswick Chap, CC BY SA 3.0

The way one species can affect the functioning and health of an entire ecosystem became evident when the US Fish and Wildlife Service reintroduced the gray wolf into Yellowstone National Park in 1995. At one time wolves populated much of the Untied States but were hunted nearly to extinction, and the last wolf in Yellowstone was killed in the 1930s. Wolves are a keystone species, which is one that is crucial to the health and structure of an ecosystem. In Yellowstone it was seen that because the elk no longer had a natural predator, they were increasing in population to the point that they were overgrazing and killing plants. The results of reintroducing the wolves to keep the elk population down were extensive and varied. Trees and bushes began to regrow, stabilizing riverbanks and providing homes for wildlife. Beavers returned to the streams and their dams slowed and stored water and added wetlands where many plants and animals thrive. The carrion left over from animals killed by wolves fed other animals such as ravens, eagles, coyotes, and bears. The whole Yellowstone ecosystem and its residents became healthier and more diverse just by restoring balance through the reintroduction of one species.

Gray Wolf at Yellowstone
Courtesy of Barry O'Neill, National Park Service (NPS)

Elk
Courtesy of MONGO

At this time much is being done to research the functions of different organisms in ecosystems. Governments and nonprofit organizations are working to preserve endangered species and restore health to ecosystems, and new ways of planning cities and towns are being developed that take natural ecosystems and their residents into consideration. With some thought and research and also a willingness for us to change our way of doing things, we can save endangered species and ensure the health of our own.

12.6 Solutions

A wind farm in Texas

In the past, people were unaware of how the burning of fossil fuels and other human activities affect Earth's air and water. With advances in knowledge gained by scientific research and the use of new technology, global environmental problems are being identified, causes of the problems are being discovered, and solutions found.

Throughout our study of geology, we have seen how scientists study each sphere of Earth individually. These scientists include geologists, biologists, physicists, chemists, ecologists, climate scientists, and many others. Earth system science is a scientific discipline that gathers information from the study of individual spheres and combines that information to look at Earth as a whole. Earth system scientists look at Earth's various structures and systems and how they work together, the changes that are occurring on Earth, and what those changes mean. The ways in which human activities affect the Earth system are also taken into consideration. For example, some Earth system scientists are involved in finding ways to safeguard people's water resources and food sources, and others study impacts on preservation of animals' habitats and marine life.

Solar panels

Courtesy of US Air Force/by Airman 1st Class Nadine Y. Barclay, Nellis AFB, Nevada

Although there are many problems within the Earth system, including carbon dioxide emissions, other types of pollution, and overuse of resources, scientists are beginning to find solutions using new approaches. Many scientists are now interested in researching sustainability, or using resources in ways that do not deplete them or damage the Earth system. Nature provides everything we need to survive and be productive and healthy. However, we in turn must protect and preserve the healthy functioning of natural systems. For example, scientists are researching ways to efficiently produce energy without creating pollution. Windmills can be used to harness energy from the winds in the atmosphere, dams can be used to collect energy from water moving in the hydrosphere, and solar panels can be used to gather energy from the Sun that is made available to us through the magnetosphere and atmosphere.

Earth system science is an exciting new area of study, and scientists are doing some interesting new research. For example, by collecting information about the hydrosphere, biosphere, and atmosphere, scientists are pursuing the idea of making sewage treatment plants that filter the dirty water through a wetlands-like environment. The idea is to use the natural functions of bacteria, algae, and plants to clean the water.

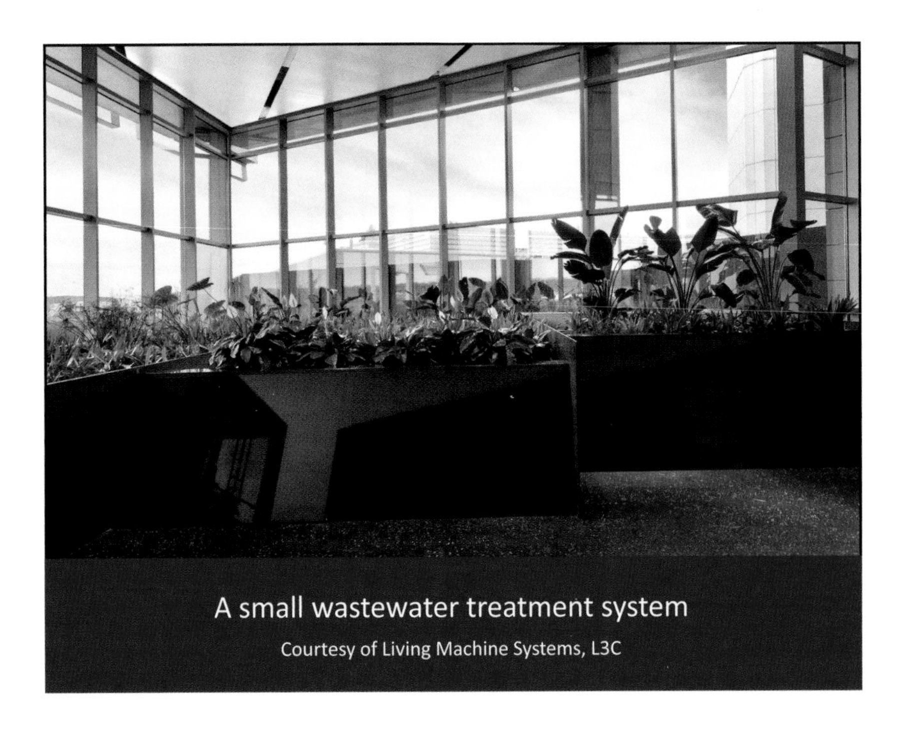

A small wastewater treatment system
Courtesy of Living Machine Systems, L3C

Scientists, engineers, and architects are also involved in designing and building green buildings that are energy efficient, use less water, reduce pollution, are partially constructed from recycled materials, and in general have less impact on the environment around them as well as being healthier workplaces. Some estimates show that as much as 24% of global carbon dioxide emissions come from buildings, mainly through heating and cooling. By using new technologies for insulation, window design, solar energy, and other ideas, carbon emissions from buildings can be significantly lowered.

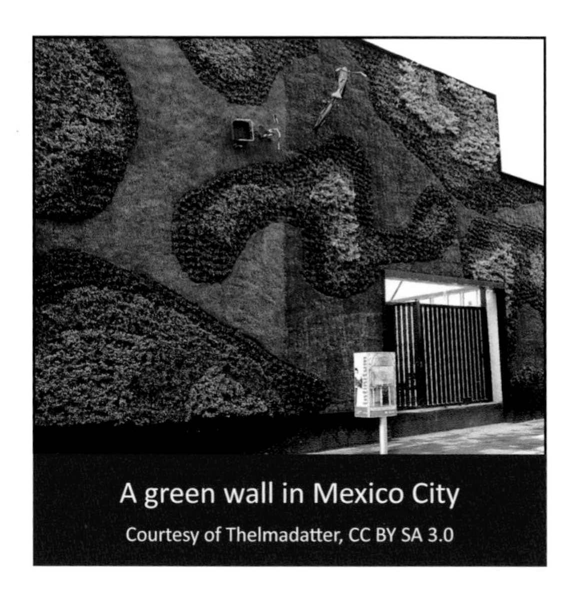
A green wall in Mexico City
Courtesy of Thelmadatter, CC BY SA 3.0

Another idea being incorporated in buildings is the green wall. Plants in vertical gardens hung on exterior walls of buildings help clean some of the CO_2 from the air, provide an insulating effect from temperature and noise, and create a habitat for birds and insects. Interior green walls can clean toxins from the air and can be used to filter and clean water that flows through them. Green walls also add beauty and calm to our environment, reducing stress.

Turkey Tail Mushroom

Sometimes interesting new research begins by looking at nature on a small scale. For example, scientists have found that mushrooms have many important functions. Some can use plastics, oil, or radioactive materials for food, and scientists are looking for ways that mushrooms might help clean our environment. In addition, research is being done to find out how mushrooms may be useful as medicines.

These are only a very few examples of research that is taking place now. Each new discovery leads to many more questions, possibilities, and new ideas. This is a very exciting time to be a scientist!

12.6 Summary

- The different spheres of Earth are interdependent. They interact with and rely on each other in order to function properly.
- A change in conditions in one sphere can cause changes to other spheres.
- Human activities can lead to changes in Earth's spheres.

- So far, the data show that the worldwide yearly average global temperature of Earth is increasing.
- Global climate change is one cause of changes to habitats that are making some species endangered.
- In a process called ocean acidification, oceans are becoming more acidic as atmospheric CO_2 levels rise.
- Earth system science is a field of study in which information from the study of Earth's individual spheres is combined to look at Earth as a whole.
- Many Earth system scientists are interested in sustainability, or finding ways to use resources in ways that do not deplete them or damage the Earth system.

12.7 Some Things to Think About

- Think of a natural event and describe how you think it would affect the different spheres of Earth.
- Think about the area where you live. Knowing that all of Earth's spheres are interconnected, how do you think you and your area might be connected to a place on the other side of the globe?
- Think of a recent natural disaster that you have heard about on the news. How did this event affect people in the area? How did it affect the different spheres of Earth? Were the effects only local, or were other areas affected as well?
- What are some problems that are considered to be caused by excessive amounts of carbon dioxide entering the atmosphere?
- How do you think the quantity of carbon dioxide emissions could be reduced? What difficulties would need to be overcome?
- What are some things you think we can do to help preserve plant and animal species and keep them healthy?
- Why do you think species preservation can be important?
- What are some environmental problems that you think need to be solved? Which of these problems would you like to work on? What ideas do you have for solving it?

Glossary-Index

[Pronunciation Key at end]

abiotic (ā-bī-'ä-tik) **factor** • a non-biological component of an ecosystem, 114

acid rain • rainfall that is made acidic due to pollution; can cause harm to plants and cause water to be polluted, 62, 100

aggregate ('a-gri-gət) • in geology, a rock made of a mixture of mineral fragments "glued" together by high heat and pressure within the Earth, 23

air pressure • see atmospheric pressure

aluminum • (ə-'lü-mə-nəm) • an element commonly found in rocks; its symbol is Al, 20, 22

anticline ('an-ti-klīn) • land that has been folded into an arch shape by the pushing together of land masses, 47

aquifer ('ak-wə-fər) • a large underground area that holds water, 98-99, 100-101

Arctic Ocean ('ärk-tik 'ō-shən) • the smallest of the five oceans; located at the northernmost part of the Earth, 90, 94

Aridisols (ə-'ri-di-sälz) • [L., *aridus,* dry; *solum,* soil] in soil taxonomy, the order for dry soils, 30-31

ash, volcanic • tiny solid particles ejected by a volcano, 50, 63, 67, 68, 69-70, 134

asthenosphere (as-'the-nə-sfir) • [*asthenes,* Gr., weak] the middle layer of the mantle; thought to have a putty-like texture, 36, 38-39, 40, 44, 45, 48, 124, 132

Atlantic Ocean (ət-'lan-tik 'ō-shən) • second largest ocean; located between the eastern coast of the Americas and the western coasts of Europe and Africa, 90, 94, 133

atmosphere ('at-mə-sfir) • [Gr., *atmos,* vapor] the layer of gases that surrounds the Earth, 56, 57, 59-64, 69, 70, 76-85, 88, 89, 100, 101, 105, 107-114, 127, 128, 132, 133-139, 144

atmospheric pressure (at-mə-'sfir-ik 'pre-shər) • air pressure; the force exerted by the weight of air molecules in Earth's atmosphere, 81-82

aurora (ə-'rôr-ə) • moving streamers or curtains of colored light in a polar region; caused by electrons colliding with oxygen and nitrogen atoms in the atmosphere, 80, 128

aurora australis (ə-'rôr-ə ô-'strā-ləs) • an aurora that occurs in the region near the South Pole; also called southern lights, 128

aurora borealis (ə-'rôr-ə bôr-ē-'a-ləs) • an aurora that occurs in the region near the North Pole; also called northern lights, 128

autotroph ('ô-tə-trōf) • organisms that make their own food by photosynthesis, 117

barometer (bə-'rä-mə-tər) • an instrument that measures air pressure, 81

biogeochemical cycle (bī-ō-jē-ə-'ke-mi-kəl 'sī-kəl) • the flow of chemical elements through biological and non-biological parts of an environment, 106

biome ('bī-ōm) • a large region that is home to plants and animals requiring a similar environment, 115-116

biosphere ('bī-ə-sfir) • [Gr., *bios* life] the part of Earth that includes all living things, 56, 58, 59-64, 105-118, 132, 134, 136, 144

biotic (bī-'ä-tik) **factor** • a biological component of an ecosystem, 114

bow (bou) **shock** • the area in space where the solar wind is suddenly slowed as it encounters the forces of the geomagnetic field, 126, 127, 129

calcite ('kal-sīt) • a non-silicate mineral found in rocks and soils, 23, 28, 29

calcium ('kal-sē-əm) • an element commonly found in rocks; its symbol is Ca, 20, 22, 29

carbon ('kär-bən) • an element that is able to form more kinds of molecules than any other element, 20, 67, 77, 109-110, 136, 139, 144

carbon cycle ('kär-bən 'sī-kəl) • a cycle in the biosphere during which carbon moves from the atmosphere through plants and animals and back into the atmosphere, 16, 60, 106, 109-110, 134-140

carbon cycle ('kär-bən 'sī-kəl), **fast** • the part of the carbon cycle in which carbon is cycled more quickly between living and nonliving components, 109-110, 136

carbon cycle ('kär-bən 'sī-kəl), **slow** • the part of the carbon cycle in which carbon is cycled more slowly between living and nonliving components and has rocks as the main reservoir, 109, 110, 136

carbon dioxide ('kär-bən dī-'äk-sīd) **(CO_2)** • a molecule made of one carbon atom and two oxygen atoms; one of the greenhouse gases, 29, 62, 63, 76, 77, 83, 84, 92, 107, 108, 109, 110, 113, 132, 136-137, 139, 143, 144

carnivore ('kär-nə-vôr) • an animal that eats other animals, 117

cellular respiration ('sel-yə-lər re-spə-'rā-shən) • the process used by living things in which carbon dioxide is given off as a by-product of the chemical reactions that turn nutrients into energy, 109, 110

chemical element ('ke-mi-kəl 'e-lə-mənt) • an atom; one of the fundamental units that make up all matter, 19, 20, 106

cinder cone volcano • see volcano, cinder cone

circulate ('sər-kyə-lāt) • to follow a circular motion, 44, 133

climate ('klī-mət) • the typical or average weather in a certain region over a long period of time, 135

climate ('klī-mət) **map** • a map that shows temperatures and amount of precipitation for an area, 11-12

climate ('klī-mət) **zone** • an area that has similar temperature ranges and precipitation, 12

CME • coronal mass ejection, 128-129

compass ('kəm-pəs) • a device that determines direction by means of a magnetic needle that is attracted to Earth's Magnetic North Pole, 11, 121-122, 124

composite volcano • see volcano, composite

conclusion (kən-'klü-zhən) • the result or outcome of reasoning, 3

condensation (kän-den-'sā-shən) • for water vapor, the process of changing from the gaseous state to liquid water, 89

condense (kən-'dens) • for water vapor, to change from the gaseous state to liquid water, 89

continental crust • see crust, continental

Continental Divide • a watershed made up of continuous mountain ranges and ridges in the Americas; runs from Alaska through the Andes Mountains; also called the Great Divide, 94

convection (kən-'vek-shən) • the transfer of heat (energy) that occurs when molecules in a liquid that is being heated bump into each other, causing movement in the liquid, 39, 40, 44, 45, 123, 124

core • the center of Earth; contains iron and nickel; made of two layers — the outer core and the inner core, 23, 34, 39, 40, 41, 44, 56, 72-73, 76, 121, 122-123, 132

core drill • a drill that cuts a circular hole as it moves downward into the Earth, creating cylindrical rock samples, 14

core, inner • the very center of the Earth; thought to be solid, 34, 40, 41, 72-73, 122-123

core, outer • the liquid outer layer of the core of the Earth, 34, 40, 41, 56, 72-73, 122-123

core sample • a cylinder of rock that is drilled from the Earth; used to test and study layers of rock, 14, 124

coronal mass ejection (kə-'rōn-nəl 'mas i-ˈjek-shən) **(CME)** • a huge cloud of plasma blasted from an area of the Sun due to a disturbance in the Sun's magnetic field, 128-129

crack hammer • a small sledgehammer with two blunt ends; used to break apart rocks, 11

crust • the outermost layer of Earth, 19, 23, 34-36, 37, 38, 44, 47, 51, 52, 56, 66, 67, 69, 70, 71, 72, 77, 91, 124

crust, continental (kän-tə-'nen-təl) • the part of the outer layer of Earth that forms the continents and other land masses, 35, 36

crust, oceanic (ō-shē-'a-nik) • the part of the outer layer of Earth that lies below the oceans, 35-36, 91

cryosphere ('krī-ō-sfir) • the part of the hydrosphere that is frozen water, 57

cycle ('sī-kəl)) • a series of events that repeats, 60, 62, 88

cycle, carbon • see carbon cycle

cycle, energy • see energy cycle

cycle, hydrologic • see hydrologic cycle

cycle, nitrogen • see nitrogen cycle

cycle, rock • see rock cycle

cycle, water • see hydrologic cycle

decomposer (dē-kəm-'pō-zər) • an organism that breaks down dead organisms, 117

deform (di-'fôrm) • in geology, to alter the shape of something by stress, 6, 37

deformation (dē-fôr-'mā-shən) • the action of deforming; the state of being deformed, 6

density ('den-sə-tē) • the amount of mass contained in a certain amount of three-dimensional space, 40

detrital sedimentary rock • see sedimentary rock, detrital

diamond ('dī-mənd) • a mineral made of pure carbon, 28, 67

dissolution (di-sə-'lü-shən) • breaking down into component parts, 99

dolomite ('dō-lə-mīt) • a non-silicate mineral found in rocks and soils, 23

dome mountain • see mountain, dome

dynamic (dī-'na-mik) • changing over time, 13, 40, 44, 129

dynamic process (dī-'na-mik 'prä-ses) • a series of physical or energetic changes that occur over time, 13, 40, 44-52

dynamo ('dī-nə-mō) • a machine that converts mechanical energy into electrical energy, 122

earthquake • a mild to severe shaking of the Earth's crust due to the buildup and release of stress caused by movement of tectonic plates, 5, 38, 44-45, 50-52, 70-72, 93

Earth system science • the scientific discipline that gathers information from the study of the individual spheres of Earth and combines that information to look at Earth as a whole, 143-145

ecological community (ē-kə-'lä-ji-kəl kə-'myü-nə-tē) • all the populations of the different habitats in an ecosystem taken together, 115, 118

ecology (i-'kä-lə-jē) • the study of the interactions and relationships of living things to the world around them, 114

ecosystem ('ē-kō-sis-təm) • a specific area that contains similar living things that exist under similar conditions, 62, 105, 113-115, 117-118, 132-133, 140, 141-142

electronic (i-lek-'trä-nik) **tool** • a tool that contains components that direct the transmission of electrons, 10, 12-14

endangered species (in-'dān-jərd 'spē-shēz) • plant, fungi, and animal species whose numbers are declining to the point where they are in danger of becoming extinct, 134, 140-143

energy cycle ('e-nər-jē 'sī-kəl) • a cycle in the biosphere in which heat and light energy from the Sun is used by plants to make food, 60, 106, 112-113

environment (in-'vī-rən-mənt) • everything that surrounds a living thing in the area where it lives, 7, 15, 64, 95, 114-115, 141, 144, 145

environmental geologist • see geologist, environmental

epicenter ('e-pi-sen-tər) • the point of origin of an earthquake, 70, 71-72

erode (i-'rōd) • to wear away through the action of rain, wind, glaciers, chemicals, and other forces, 60, 91, 97

erosion (i-'rō-zhən) • the process or result of eroding, 6, 56, 106, 114

eruption (i-'rəp-shən) • in geology, the process of forcing magma through the Earth's crust to the surface, 15, 44, 48-50, 56, 63, 67, 68, 77, 93, 110, 134

escape velocity (is-'kāp və-'lä-sə-tē) • the minimum speed needed for an object to escape from the gravitational field of a star, planet, or moon, 82

evaporate (i-'va-pə-rāt) • to change water into its gaseous state, 88, 101

evaporation (i-va-pə-'rā-shən) • the process of evaporating, 25, 90

exchange pool • the part of a biogeochemcial cycle where elements and molecules are cycled relatively quickly between living and nonliving components, 106, 107, 109, 111

exosphere ('ek-sō-sfir) • the outermost layer of Earth's atmosphere; is sometimes considered to be part of outer space, 78-79, 81

fast carbon cycle • see carbon cycle, fast

fault • a fracture in the surface of the Earth, 47, 51-52, 71

fault-block mountain • see mountain, fault-block

fault line • the area where the two sides of a fracture in the surface of the Earth meet, 51-52

feldspar ('feld-spär) • a common silicate mineral that contains silicon, oxygen, aluminum, and either potassium, calcium, or sodium, 22, 24

folded mountain • see mountain, folded

foliated ('fō-lē-ā-təd) • [*foliare,* L., leaf) layered, 26

food chain • the order in which particular organisms in an ecosystem feed on each other, 115, 117-118, 141

food web • a system of interrelated food chains in an ecosystem; describes the interrelationships of how species feed on and are fed by each other and tracks how energy moves through the organisms in an ecosystem, 117-118, 140

foraminifera (fə-ra-mə-'ni-fə-rə) • a type of zooplankton, 139-140

formulate ('fôr-myə-lāt) • to put data into an ordered statement, 3

Fourier ('fô-rē-ā), **Jean-Baptiste Joseph** (zhän-bap-'tēst 'jō-zəf) • (1768-1830 CE), a French mathematician and physicist who first proposed the theory of the greenhouse effect, 136

fracture ('frak-chər , -shər) • in geology, a break or crack in rock, 48, 51, 97, 98, 106

garnet ('gär-nət) • a mineral, 67, 69

geochemistry (jē-ō-'ke-mə-strē) • the study of the chemistry of Earth, 6

Geographic (jē-ə-'gra-fik) **North Pole** • the northernmost point of Earth's axis of rotation; also called true north, 121

Geographic (jē-ə-'gra-fik) **South Pole** • the southernmost point of Earth's axis of rotation, 121

geologist (jē-'ä-lə-jist) • a scientist who studies the Earth, 5

geologist, environmental (jē-'ä-lə-jist, in-vī-rən-'mən-təl) • a scientist who studies environmental changes caused by human activities, 7

geology (jē-'ä-lə-jē) • [Gr., *geo,* earth; *logy,* the study of] the study of Earth, 2

geology, historical (jē-'ä-lə-jē, hi-'stôr-i-kəl) • the branch of geology that uses science to examine how Earth came into being and how it has changed, 2, 4

geology, physical (jē-'ä-lə-jē, 'fi-zi-kəl) • the branch of geology that examines Earth's processes and the chemical and physical nature of Earth, 2

geology, resource • the branch of geology that studies natural resources used by humans, 6

geology, structural • the branch of geology that deals with the internal structure, form, and arrangement of rocks, 6

geomagnetic (jē-ō-mag-'ne-tik) **field** • the region surrounding Earth in which magnetic forces can be observed, 122-127

geomagnetic reversal (jē-ō-mag-'ne-tik ri-'vər-səl) • a shift in Earth's geomagnetic field in which the Magnetic North Pole and the Magnetic South Pole switch locations, with north becoming south and south becoming north, 123-126

geomagnetic (jē-ō-mag-'ne-tik) **storm** • a disturbance in Earth's magnetosphere caused by a large burst of radiation from the Sun, 128-129

geosphere ('jē-ō-sfir) • the rock part of Earth; includes the crust, mantle, and core, 56, 59-64, 66-73, 105

glacier ('glā-shər) • a large body of ice and snow that moves very slowly down sloped land, 57, 58, 61, 84, 88, 95-96, 106, 134

global climate change • global changes in different aspects of the climate, including weather patterns, average global temperature, etc., 135-145

global conveyor belt • continuous ocean currents that travel around the world, 92-93

global positioning system • see GPS

gneiss ('nīs) • a metamorphic rock; commonly ejected by volcanoes, 67

GPR (ground penetrating radar) • a method that uses high frequency radio waves to image below the surface, 13

GPS (global positioning system) • a device that uses satellite information to provide the location and time on or near the Earth, 13

granite ('gra-nət) • an igneous rock composed mainly of quartz and feldspar, 23-24, 29, 35, 67, 97

Great Divide • the Continental Divide, 94

green building • a structure that is designed to have as small an environmental impact as possible; also the processes involved in planning, designing, constructing, and operating such a building, 144-145

greenhouse effect •the trapping of energy from the Sun by Earth's atmosphere, 62, 83-85, 113, 136

greenhouse gas • a gas in the atmosphere that traps heat from the Sun, 63-64, 83-85, 113, 136-137

green wall • a vertical garden hung on an exterior or interior wall of a building, 145

ground penetrating radar • see GPR

groundwater • water that is found in the soil and rocks below the surface of the land, 13, 15, 57, 88, 89, 96-99, 100

gypsum ('jip-səm) • a non-silicate mineral found in rocks and soils, 23, 28

habitat ('ha-bə-tat) • an area that contains the set of conditions that a specific type of organism needs in order to live and reproduce, 5, 56, 60, 115, 117, 134, 141, 143, 145

halite ('ha-līt , 'hā-) • a non-silicate mineral found in rocks and soils, 23

hammer, crack • see crack hammer

hammer, rock • see rock hammer

handpick • a rock hammer

hand tool • a tool held in the hand and operated without electricity, 10-12

hematite ('hē-mə-tīt) • an iron-containing mineral that can look gray but leaves a red streak, 28-29

herbivore ('ər-bə-vôr) • an animal that eats plants, 117

historical geology • see geology, historical

hydrologic cycle (hī-drə-'lä-jik 'sī-kəl) • the continuous movement of water through the various water systems on Earth; also called the water cycle, 16, 60, 88-90, 99, 101

hydrologist (hī-'drä-lə-jist) • a scientist who studies water and the hydrosphere, 88

hydrosphere ('hī-drə-sfir) • [Gr., *hydro,* water] all the water surrounding Earth, 56, 57, 59-64, 88-102, 105, 107, 109, 114, 132-134, 136, 144

igneous ('ig-nē-əs) **rock** • a rock formed from molten magma; the most plentiful of the three basic types of rock, 23, 24, 25, 26, 27, 35, 36, 67, 97

inches of mercury ('mər-kyə-rē) • unit of measure of atmospheric pressure, 81

Indian Ocean ('in-dē-ən 'ō-shən) • third largest ocean; located between the eastern coast of Africa and the western coast of Australia, 90

infiltrate ('in-fil-trāt) • to pass into or through a substance by means of pores or void spaces, 89, 97, 98, 100

inner core • see core, inner

inorganic (in-ôr-'ga-nik) • a substance that does not contain carbon atoms, 19, 20

interdependent (in-tər-di-'pen-dənt) • one or more things that interact with and rely, or depend, on each other, 111, 114, 132

interpretation (in-tər-prə-'tā-shən) • explanation, 3

iron • an element commonly found in rocks; its symbol is Fe, 20, 22, 26, 28, 40, 73, 122-123

keystone species ('spē-shēz) • a species that is crucial to the health and structure of the ecosystem it lives in, 142

lake • a large body of surface water surrounded by land, 56, 57, 88, 89, 93, 94, 95, 96, 97, 99, 100, 116

lava ('lä-və) • molten rock that has reached the surface of the Earth, 48-50, 63, 67-68, 72, 124, 134

limestone ('līm-stōn) • a type of chemical sedimentary rock; under heat and pressure, turns to marble, 25, 26, 29, 98, 99

lithosphere ('li-thə-sfir) • [Gr., *lithos*, stone] the rigid upper layer of the mantle, 36, 37, 38, 44, 48, 70, 106, 109, 110, 114, 132, 134, 136

magma ('mag-mə) • rocks that have melted due to high temperatures within the Earth, 23, 26, 36, 47, 48, 66, 67, 69, 71, 97, 124

magnesium (mag-'nē-zē-əm) • an element commonly found in rocks; its symbol is Mg, 20, 22

Magnetic North Pole • the pole of Earth's magnetic field toward which a compass needle points, 121-122, 123-124, 128

Magnetic South Pole • the location on the surface of Earth in the Southern Hemisphere that is southern end of Earth's magnetic field, 121-122, 124, 128

magnetometer (mag-nə-'tä-mə-tər) • an instrument that measures magnetic force, 124

magnetopause (mag-'nē-tə-pôz) • the outer boundary of the magnetosphere, 127

magnetosheath (mag-'nē-tō-shēth) • the transition area between the bow shock and the magnetosphere, 126-127

magnetosphere (mag-'nē-tə-sfir) • the area surrounding Earth that contains Earth's magnetic field and is created by the interaction of Earth's magnetic field with gases from the Sun, 15, 56, 58, 59, 63, 68, 114, 121-129, 132, 133, 144

magnetotail (mag-'nē-tō-tāl) • the part of the magnetosphere that extends far beyond Earth on the night side, 126-127, 128, 129

mantle ('man-təl) • the layer of the Earth that is just below the crust; contains 3 layers — lithosphere, asthenosphere, and mesosphere, 34, 36-40, 41, 44, 48, 56, 69, 72, 132

map • a diagram showing the different parts of an area, e.g., roads, cities, mountains, 11-12, 45, 124

map, climate • see climate map

map, topographic • see topographic map

marble • metamorphic rock formed from limestone that has been exposed to heat and pressure, 26

marsh • a wetland that typically contains grasses, reeds, and some open water, 93, 94, 95

mass • commonly, the property of matter that gives it weight, 40, 78, 82

mesopause ('me-zō-pôz) • the upper boundary of the mesosphere in the atmosphere, 78-79

mesosphere ('me-zō-sfir) • [Gr., *mesos,* middle or between]:

1. the deepest layer of the mantle; located between the core and the asthenosphere, 36, 39-40, 41
2. the layer of Earth's atmosphere that is above the stratosphere, 78-80

metamorphic (me-tə-'môr-fik) **rock** • [Gr., *meta,* along, with, or between; *morphe,* shape] rock formed when one rock type transforms into another due to heat, pressure, or chemical reactions; one of the three basic types of rock, 23, 25-26, 27, 35, 67

metamorphosis (me-tə-'môr-fə-səs) • [Gr., *meta,* along, with, or between; *morphe,* shape] the transformation or change of one thing into another, 25

mica ('mī-kə) • a very soft, layered silicate mineral; contains oxygen, silicon, and either potassium, magnesium, iron, or aluminum, 22, 26

mid-oceanic ridge • a chain of mountains under the ocean that is formed by seafloor spreading, 124

millibar (mb) ('mi-lə-bär) • a unit of measure of atmospheric pressure, 81-82

mineral ('min-rəl) • a material that is naturally formed, is a solid, is inorganic, and has an internal structure or organization of atoms, 6, 14-15, 19, 20-23, 25, 26, 27-29, 38, 48, 56, 66, 67, 91

mineral, rock forming • a mineral found in rocks, 20

Mohs, Friedrich (mōz, frē-drik) • [1773-1839 CE] German geologist; developed the Mohs scale of mineral hardness for scratch testing minerals, 27

Mohs (mōz) **scale of mineral hardness** • a chart of the relative hardness of 10 minerals and other objects to be used in scratch testing mineral samples, 27-29

Mollisols ('mä-lə-sälz) • [L., *molere,* to grind or crush; *solum,* soil] in soil taxonomy, the order for prairie soils obtained from parent materials that have been ground or crushed, 30-31

molten ('mōl-tən) • melted, liquefied, 23, 37, 44, 47, 48, 56, 66, 72, 122, 123, 124

mountain, dome • a mountain formed when magma pushes up under the crust, creating a large raised domed shape on the Earth's surface, 47

mountain, fault-block • a mountain formed by plates sliding up and down with respect to each other, 47

mountain, folded • a mountain formed by the movement of plates, which causes land masses to be pushed together, folding them, 47

niche ('nich *or* 'nēsh) • the role a species plays within an ecosystem, 115

nitrogen ('nī-trə-jən) • an element that makes up 78% of the atmosphere, 60, 76, 106, 111-112, 127, 128, 132

nitrogen cycle ('nī-trə-jən 'sī-kəl) • a cycle in the biosphere in which nitrogen moves from the atmosphere, through plants and animals, and back into the atmosphere, 60, 106, 111-112

nitrogen fixation ('nī-trə-jən fik-'sā-shən) • the process during which bacteria in the soil convert nitrogen to a molecule plants can use, 111-112

northern lights • an aurora that occurs in the region near the North Pole; also called aurora borealis; 80, 128

ocean ('ō-shən) • any of the bodies of salt water that cover 71% of Earth's surface, 15, 19, 36, 45, 57, 58, 77, 84, 88-93, 94, 95, 96, 99, 100, 101, 102, 105, 110, 112, 115, 116, 124, 132, 133, 137, 138, 139-140

ocean acidification ('ō-shən a-sə-də-fə-'kā-shən) • the process by which seawater becomes more acid; the increasing acidity of seawater, 139-140

oceanic crust •see crust, oceanic

olivine ('ä-lə-vēn) • a mineral, 67

organic (ôr-'ga-nik) • a substance that contains carbon atoms, 19, 20, 62,105-106, 108, 110, 136

outer core • see core, outer

oxygen ('äk-si-jən) • an element commonly found in rocks and as a gas in the atmosphere; its symbol is O, 20-22, 63, 76-78, 81-82, 84, 91, 92, 106, 107-109, 127, 128, 132

oxygen cycle ('äk-si-jən 'sī-kəl) • the cycle that moves oxygen through the biosphere, 106, 107-109

ozone ('ō-zōn) **(O_3)** • a molecule made of three oxygens; found in the atmosphere, 76, 77, 79-80, 109, 112

ozone ('ō-zōn) **layer** • the atmospheric layer in the lower stratosphere that absorbs harmful ultraviolet radiation, 76, 77, 79-80, 112

Pacific Ocean (pə-'si-fik 'ō-shən) • the largest ocean; stretches between the west coast of the Americas to the eastern coast of Asia and Australia, 90, 94

paleomagnetism (pā-lē-ō-'mag-nə-ti-zəm) • the study of magnetism in ancient rocks to learn about Earth's geomagnetic field, 125-126

paleomagnetist (pā-lē-ō-'mag-nə-tist) • a scientist who studies magnetism in ancient rocks to learn about Earth's geomagnetic field, 125-126

paleontologist (pā-lē-än-'tä-lə-jist) • a scientist who studies the fossil records contained in ancient rocks, 125-126

parent material • the source of the weathered matter from which soils develop, 29-30

peridotite (pə-'ri-də-tīt) • an igneous rock made mostly of the minerals olivine and pyroxene, 67, 69

permeable ('pər-mē-ə-bəl) • having pores that are connected, allowing water to flow from pore to pore through rock or soil, 97, 99

photosynthesis (fō-tō-'sin-thə-səs) • the process used by plants in which carbon dioxide, sunlight, and water are converted to sugars to use as food, and oxygen is released as a waste product, 77, 84, 107-110, 112, 117, 132, 134

photosynthetic (fō-tō-sin-'the-tik) • relating to the process of photosynthesis, 107

physical geology • see geology, physical

phytoplankton (fī-tō-'plank-tən) • a group of microscopic organisms found in both fresh and salt water; includes bacteria, protists, and single-celled plants, 107-108, 110, 112, 140

plasma ('plaz-mə) • a state of matter containing electrically charged particles (negatively charged electrons and positively charged protons and ions), 126-129

plastic • pliable; able to be molded, 38

plate • in geology, a slowly moving, rigid piece of the lithosphere that contains land masses or ocean floor, 37, 38, 44-46, 47, 51-52, 71

plate tectonics (tek-'tä-niks) • the theory that earthquakes, volcanoes, and mountain formation are caused by the movement of pieces of Earth's lithosphere, 44-52

pollute (pə-'loot) • to contaminate with chemicals or other waste products, 99, 100, 141

pollution (pə-'lü-shən) • the action of polluting; the condition of being polluted, 15, 62, 94, 100, 102, 143, 144

soil • the upper layer of Earth's surface; made of weathered rocks and the remains of plants and animals; dirt, 13, 16, 19, 20, 22, 23, 29-31, 34, 35, 55, 56, 60, 62, 63, 66, 88, 96, 97, 100, 101, 105, 106, 111, 114, 115, 132

soil order • in soil taxonomy, the main category for classifying soils; there are twelve soil orders, 30-31

soil taxonomy (tak-'sä-nə-mē) • a system for categorizing soils based on their properties, 29-31

solar storm • an intense burst of radiation from the Sun caused by disturbances in the Sun's magnetic field, 128

solar wind • the massive stream of plasma coming from the Sun, 58, 126-127, 129

southern lights • an aurora that occurs in the region near the South Pole; also called aurora australis; 80, 128

Southern Ocean ('sə-thərn 'ō-shən) • the fourth largest ocean, located near Antarctica, 90

sphere ('sfir) • a ball shaped object; in geology, one of the interconnected layers that make up the Earth, 37, 38, 56-64, 85, 88, 105, 106, 132-136, 141, 143

static • staying in once place; not moving or changing, 13

stratopause ('stra-tə-pôz) • the upper boundary of the stratosphere, 78-79

stratosphere ('stra-tə-sfir) • the layer of the Earth's atmosphere that is above the troposphere, 78-79

stratovolcano (stra-tō-väl-'kā-nō) • a volcano formed from alternating layers of lava flow and lava rubble; also called a composite volcano, 50

streak plate • an unglazed ceramic tile used to show the color of a rock or mineral sample when the sample is rubbed across it, 27-29

stress • a force resulting from pressure or tension, 51, 71

structural geology • see geology, structural

sublimation (sə-blə-'mā-shən) • the process of matter changing from a solid state to a gaseous state without first passing through the liquid state, 89

surface water • water found on the land, such as lakes, rivers, swamps, and marshes, 93-96, 97

sustainability (sə-stā-nə-'bi-lə-tē) • a way of using resources so they are not depleted and damage is not done to the Earth system, 144

swamp • a wetland that typically contains trees, 29, 93, 94, 95

symbiotic (sim-bē-'ä-tik) • having an interdependent relationship, 111

syncline ('sin-klīn) • land that has been folded into a trough shape by the pushing together of land masses, 47

system ('sis-təm) • a group of conditions, events, etc. that interact to form a whole, 56, 60, 63, 132-145

tectonics, plate • see plate tectonics

test kit, rock and mineral • see rock and mineral test kit

tetrahedral (te-trə-'hē-drəl) • relating to or having the shape of a tetrahedron, 20

tetrahedron (te-trə-'hē-drən) • a solid geometric shape that has four sides, 20

theory ('thē-ə-rē) • in science, a fact-based explanation of an aspect of the natural world that is assumed to be true but is not proven, 3

thermocline ('thər-mə-klīn) • [Gr., *herme,* hot; *kline,* bed] the thin layer of water in the ocean that separates the warmer, upper layer from the colder, deeper layer, 91

thermopause ('thər-mə-pôz) • the upper boundary of the thermosphere, 78-79

thermosphere ('thər-mə-sfir) • the layer of Earth's atmosphere that is above the mesosphere, 78-80

tide • the periodic rise and fall of an ocean, 93

topographic (tä-pə-'gra-fik) **map** • a map with contour lines that show the shape and elevation of an area, 11

topography (tə-'pä-grə-fē) • [Gr., *topos,* place; *graphe,* to write] features of a landscape, such as steepness and height, 29

trophic ('trō-fik) **level** • an organism's position on the food chain or in a food web, 117-118, 140

tropopause ('trō-pə-pôz) • the upper boundary of the troposphere, 78-79

troposphere ('trō-pə-sfir) • the layer of the atmosphere that is closest to the Earth's surface, 78-79

true north • the Geographic North Pole, 121

tsunami [(t)soo-'nä-mē] • an immense wave caused by a large quantity of water that is suddenly moved by an underwater earthquake, volcanic eruption, or landslide, 44, 93, 134

ultraviolet radiation (əl-trə-'vī-ə-lət rā-dē-'ā-shən) • energy from the Sun that is invisible to the eyes and can be harmful to organisms, 77, 79, 80, 109, 112

vent • in geology, an opening in the ground that lava comes through, 49

viscosity (vis-'kä-sə-tē) • a measure of how resistant a substance is to flowing; how thick or thin a fluid is, 48, 49

volcano (väl-'kā-nō) • a mountain or hill formed by magma that has been forced through a break in the Earth's crust, 5, 29, 36, 40, 44-45, 48-50, 56, 67-70, 72, 91, 105, 106, 132, 134

volcano, cinder cone (väl-'kā-nō, 'sin-dər 'kōn) • a small volcano with steep sides that forms around a vent, 49

volcano, composite (väl-'kā-nō, käm-'pä-zət) • a volcano formed from alternating layers of lava flow and lava rubble; also called a stratovolcano, 50

volcano, shield (väl-'kā-nō, 'shēld) • a volcano with long, sloping sides formed from layers of low viscosity lava, 49

volume ('väl-yüm) • the amount of a substance occupying a three-dimensional space, 40, 76, 81, 97, 138

water cycle • see hydrologic cycle

watershed • the area of land where all the precipitation that falls on it flows, or drains, into a common body of water such as a stream, river, or lake, 93-94

water vapor ('vā-pər) • the gaseous state of water, 57, 76, 77, 79, 83, 88-90, 100, 113

weather:

1. with regard to rocks, to wear away, 24, 29, 98, 106
2. with regard to a set of conditions in the atmosphere, how hot or cold, windy or calm, wet or dry, clear or cloudy it is, 15, 16, 57, 61, 77, 78, 79, 90, 113, 114, 135, 138

worldview • the group of philosophies and beliefs that someone uses to understand the world around them, 3-4

zooplankton (zō-ə-'plank-tən) • tiny, floating marine animals that are a major food source at the base of the ocean food web, 139, 140

Pronunciation Key

a	add	l	love	u	up
ā	race	m	move	ü	sue
ä	palm	n	nice	v	vase
â(r)	air	ng	sing	w	way
b	bat	o	odd	y	yarn
ch	check	ō	open	z	zebra
d	dog	ô	jaw	ə	a in above
e	end	oi	oil		e in sicken
ē	tree	oo	pool		i in possible
f	fit	ou	out		o in melon
g	go	p	pit		u in circus
h	hope	r	run		
i	it	s	sea		
ī	ice	sh	sure		
j	joy	t	take		
k	cool				

More REAL SCIENCE-4-KIDS Books
by Rebecca W. Keller, PhD

Building Blocks Series yearlong study program — each Student Textbook has accompanying Laboratory Notebook, Teacher's Manual, Lesson Plan, Study Notebook, Quizzes, and Graphics Package

Exploring the Building Blocks of Science Book K (Activity Book)
Exploring the Building Blocks of Science Book 1
Exploring the Building Blocks of Science Book 2
Exploring the Building Blocks of Science Book 3
Exploring the Building Blocks of Science Book 4
Exploring the Building Blocks of Science Book 5
Exploring the Building Blocks of Science Book 6
Exploring the Building Blocks of Science Book 7
Exploring the Building Blocks of Science Book 8

Focus Series unit study program — each title has a Student Textbook with accompanying Laboratory Notebook, Teacher's Manual, Lesson Plan, Study Notebook, Quizzes, and Graphics Package

Focus On Elementary Chemistry
Focus On Elementary Biology
Focus On Elementary Physics
Focus On Elementary Geology
Focus On Elementary Astronomy

Focus On Middle School Chemistry
Focus On Middle School Biology
Focus On Middle School Physics
Focus On Middle School Geology
Focus On Middle School Astronomy

Focus On High School Chemistry

Super Simple Science Experiments

21 Super Simple Chemistry Experiments
21 Super Simple Biology Experiments
21 Super Simple Physics Experiments
21 Super Simple Geology Experiments
21 Super Simple Astronomy Experiments
101 Super Simple Science Experiments

Note: A few titles may still be in production.

Gravitas Publications Inc.
www.gravitaspublications.com
www.realscience4kids.com